Practical Electronics and Electronic Engineering

Practical Electronics and Electronic Engineering

Rick Jacobs

Larsen & Keller
www.larsen-keller.com

Practical Electronics and Electronic Engineering
Rick Jacobs
ISBN: 978-1-64172-427-2 (Hardback)

Larsen & Keller
Published by Larsen and Keller Education,
5 Penn Plaza,
19th Floor,
New York, NY 10001, USA

Cataloging-in-Publication Data

Practical electronics and electronic engineering / Rick Jacobs.
p. cm.
Includes bibliographical references and index.
ISBN 978-1-64172-427-2
1. Electronics. 2. Industrial electronics. 3. Electronic apparatus and appliances.
4. Engineering. I. Jacobs, Rick.
TK7835 .P73 2020
621.381--dc23

Printed and bound in China.

For more information regarding Larsen and Keller Education and its products, please visit the publisher's website www.larsen-keller.com

Table of Contents

Preface

This book aims to help a broader range of students by exploring a wide variety of significant topics related to this discipline. It will help students in achieving a higher level of understanding of the subject and excel in their respective fields. This book would not have been possible without the unwavered support of my senior professors who took out the time to provide me feedback and help me with the process. I would also like to thank my family for their patience and support.

The technology and applications which deal with the flow, emission and control of electrons in vacuum and matter is known as electronics. The discipline of electrical engineering which is concerned with the designing of electronic circuits, devices, VLSI devices and their systems is referred to as electronic engineering. It utilizes the active and non-linear electrical components including semiconductor devices like transistors, diodes and integrated circuits. The passive electrical components that are based on the printed circuit boards are also designed in this field. There are primarily two types of circuits which are studied within this field, namely, analog circuits and digital circuits. Electronic engineering encompasses various sub-fields such as consumer electronics, digital electronics, embedded systems, analog electronics and power electronics. This book is a compilation of chapters that discuss the most vital concepts in the field of electronic engineering. Some of the diverse topics covered in it address the varied branches that fall under this category. This book is an essential guide for both academicians and those who wish to pursue this discipline further.

A brief overview of the book contents is provided below:

Chapter - What is Electronic Engineering?

The discipline of electrical engineering which makes use of the nonlinear and active electrical components in order to design electronic circuits, VLSI devices and their systems is referred to as electronic engineering. This is an introductory chapter which will introduce briefly all the significant aspects of electronic engineering.

Chapter - Analog Electronics

The electronic systems with a continuously variable signal are referred to as analog electronics. A few examples of devices which involve the use of analog electronics are voltage amplifiers, variable-gain amplifiers, buffer amplifiers, current amplifiers, power amplifiers, etc. The chapter closely examines these key examples of analog electronics to provide an extensive understanding of the subject.

Chapter - Digital Electronics

The electronics that operate on digital signals are known as digital electronics. They are less susceptible to noise and signal attenuation. A few devices which use digital electronics are multiplexers and demultiplexers. All the diverse principles of digital electronics have been carefully analyzed in this chapter.

Chapter – Semiconductors

The material that has an electrical conductivity value falling between that of a conductor and an insulator is referred to as a semiconductor. The main types of semiconductor are n-type semiconductors and p-type semiconductors. The chapter closely examines these key types of semiconductors to provide an extensive understanding of the subject.

Chapter – Diodes: Types and Applications

A two-terminal electronic component that primarily conducts current in one direction is known as a diode. The main types of diodes are Zener diode, Schottky diode, tunnel diode, varactor diodes and Gunn diode. The diverse applications of these types of diodes have been thoroughly discussed in this chapter.

Rick Jacobs

1

What is Electronic Engineering?

The discipline of electrical engineering which makes use of the nonlinear and active electrical components in order to design electronic circuits, VLSI devices and their systems is referred to as electronic engineering. This is an introductory chapter which will introduce briefly all the significant aspects of electronic engineering.

Electronics and its Application

Electronics is an applied form of science that deals with electrons. It handles electric circuits containing active elements, passive elements and other underlying techniques making it as an important part of engineering. The world is growing at a fast rate and it is relevant for the technology enthusiast to upgrade with latest changes happening in the society. Moreover, it is difficult to spend few hours without electronics gadgets and they had become an important part of our everyday routine.

- Vacuum Diode – Invented by John Ambrose Fleming: Electronics era came into existence with the invention of vacuum diode in the year 1897. Vacuum tube essentially consists two electrode plates i.e. anode and cathode. This allows current direction in one direction. But this diode is not able to control the current flow. In 1906 Lee De Forest modified and named it as Vacuum Triode. The Triode has 3 terminals. Anode, Cathode and Grid terminal. Grid electrode controls the current flow from anode to cathode. This feature could boost the power of electrical signals. Likewise, Tetrode and Pentode had evolved.

- Transistor – Invented by John Bardeen, Walter Brattain and William Shockley: The real development started with the invention of the transistor in 1948 in Bell Laboratories. Large Bulky Vacuum diodes are replaced with junction transistor. Transistors are initially made with germanium material, later on, silicon BJT (Bipolar Junction Transistor) are grown up. Most of the devices developed today are made up of silicon only due to its low cost.

- IC (Integrated Circuit) – Jack Kilby: To reduce the size and cost of the entire circuit Jack Kilby introduced a new concept. This idea entirely changed the world. The complete interconnected circuit is placed on a single chip commonly called VLSI (Very Large

Scale Integrated). Computer processors used today are made up of billions of transistors integrated on a single IC.

Need of Electronics

We are living in an electronic era where machine robots are capable to do human work with more ease and high efficiency. Capsules and tablets contain wireless sensors that collect information from the body to diagnose. Transparent smartphones will exist in the coming days, we can see through them and they may lead to the use of windows or mirrors in our home to be used as PC screens and TV monitors. Sensors are placed on the plants to detect the shortage of water and alert the farmers.

Not only above applications, there are numerous electronic applications that change our daily lives in the nearby future.

Electronic devices are made up of active and passive elements and smaller IC memories. The ICs, diodes, and transistor are made of semiconductor materials and their working is dependent on current flow through them.

Applications of Electronics

Consumer Electronics

This industry is most applicable to the common people. Consumer Electronics are devices and equipment meant for everyday use. This is again categorized as:

- Office Gadgets such as calculators, Personal computers, Scanners and Printers, FAX machine, Front Projector etc.
- Home appliances such as Washing Machine, Refrigerator, Air Conditioner, Microwave Oven, Vacuum Cleaner etc.
- Audio and Video Systems such as Headphone, VCRs, DVD players, Color TVs, Microphone and Loudspeaker, Video game consoles.
- Advanced Consumer Devices such as Setup Box, ATM, Dishwasher, Smart Phones, PDA (personal digital assistant), Barcode Scanners, POS terminals.

- Storage Devices for optical playback and taping, and portable infotainment. Examples are DVDs, HDD jukebox, Portable MP3 player.

Industrial Electronics

This industry is powerful in making real-time automation. This include: Industrial automation and motion control, Machine learning, motor drive control, Mechatronics and robotics, Power converting technologies, Photo voltaic systems, Renewable energy applications, Power electronics, and Biomechanics.

Smart Grid Systems

Smart electric systems collect information from the communication technology and react accordingly based on power consumption. It is an application of intelligence, computing, and networked electricity systems. Example: Smart Meters.

Features of Smart Grid

- Digital system based on two-way communication;
- Interactive Sensors;
- Self-monitoring and Debugging;
- Valid Distribution of electricity.

Industrial Automation and Motion Control

Machines are replacing humans these days with increased productivity, time and cost. Moreover, safety is also considered for unmanageable works. Hence to delegate the human's, automation has become the preferable choice for industries.

Image Processing

3D world has been evolved from a single or multiple 2D images. Various algorithms are developed to extract 3D information from 2D patterns.

Moreover, image processing has involved in computer graphics, Artificial intelligence, Robotics for navigation, Inspection and Assembly, Computer Vision for Face and gesture recognition, Virtual reality, medical analysis.

Medical Applications

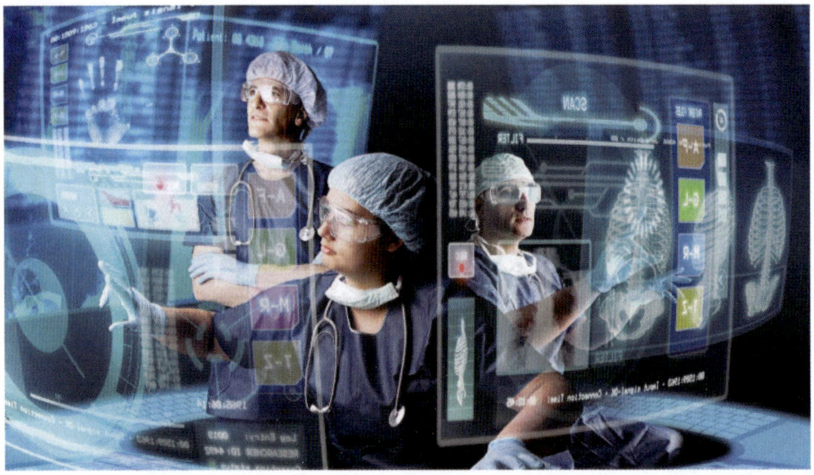

Advanced sophisticated instruments are being developed for data recording and physiological analysis. They are proven to be more useful in diagnosing diseases and for healing purpose.

Some of the medical devices and equipment used are:

- Stethoscope to listen inner sounds happening inside the human or animal body.
- Respiration Monitors for knowing the patient condition due to change in body temperature, pulse, respiration and blood flow.
- Defibrillator causes electrical shock to heart muscles and brings backs the heart to the normal working condition.
- Glucose meter for measuring sugar levels in the blood.
- Pace Maker for reducing and increasing the count of the heart beat.

Meteorological and Oceanographic

Environmental monitoring is done through various sensors and Automatic weather stations. Some of them are:

- Barometer for predicting the weather is good or bad. It gives atmospheric pressure level as an indication. If pressure level is high weather is good and if pressure is low it means the weather is bad.
- Anemometer measures the wind speed and wind direction.
- Tipping Bucket Rain gauge for measuring the rainfall periodically stored in the Automatic weather station such as Datalogger.

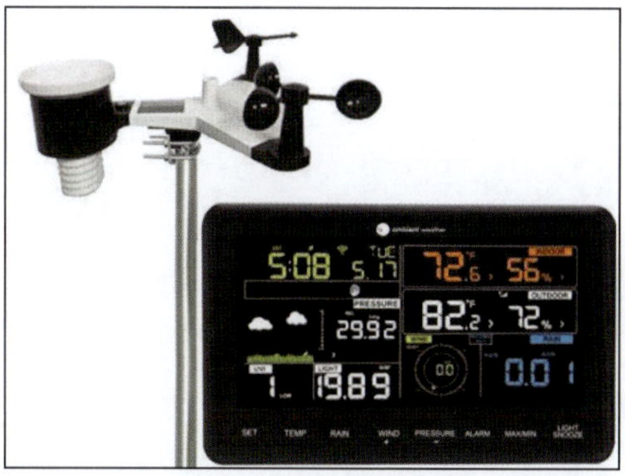

- Hygrometer measuring humidity.
- Drifter Buoy measures current, temperature, and pressure levels in the ocean.

- Data logger for storing the data collected from various sensors like humidity, temperature, wind speed and direction, solar radiation, Rain scale.

Defence and Aerospace

Defence and Aeronautical applications include:

- Missile Launching systems;
- Rocket Launchers for space;
- Aircraft systems;

- Cockpit controllers,
- Military Radars,
- Boom barrier for military applications.

Automotive (Automobiles)

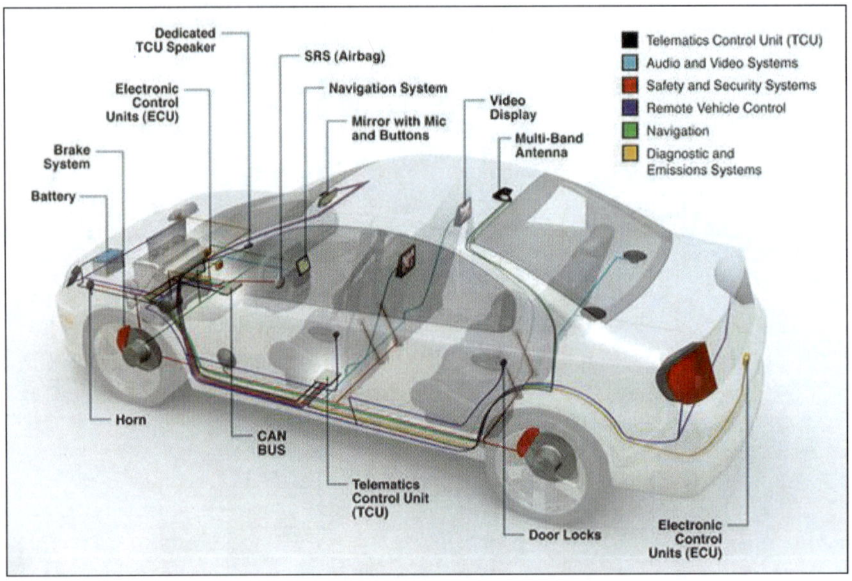

- Anti-collision unit;
- Infotainment console;
- Anti-lock braking system;
- Cruise control;
- Traction control;
- Window regulators;
- Electronic Control Unit (ECU);
- Airbag control.

Electronic Engineering

Electronic engineering (also called electronics and communications engineering) is an electrical engineering discipline which utilizes nonlinear and active electrical components (such as semiconductor devices, especially transistors, diodes and integrated circuits) to design electronic circuits, devices, VLSI devices and their systems. The discipline typically also designs passive electrical components, usually based on printed circuit boards.

Printed circuit board.

Electronics is a subfield within the wider electrical engineering academic subject but denotes a broad engineering field that covers subfields such as analog electronics, digital electronics, consumer electronics, embedded systems and power electronics. Electronics engineering deals with implementation of applications, principles and algorithms developed within many related fields, for example solid-state physics, radio engineering, telecommunications, control systems, signal processing, systems engineering, computer engineering, instrumentation engineering, electric power control, robotics, and many others.

The Institute of Electrical and Electronics Engineers (IEEE) is one of the most important and influential organizations for electronics engineers based in the US. On an international level, the International Electrotechnical Commission (IEC) prepares standards for electronic engineering, developed through consensus and thanks to the work of 20,000 experts from 172 countries worldwide.

Relationship to Electrical Engineering

Electronics is a subfield within the wider electrical engineering academic subject. An academic degree with a major in electronics engineering can be acquired from some universities, while other universities use electrical engineering as the subject. The term electrical engineer is still used in the academic world to include electronic engineers. However, some people consider the term 'electrical engineer' should be reserved for those having specialized in power and heavy current or high voltage engineering, while others consider that power is just one subset of electrical engineering, as well as 'electrical distribution engineering'. The term 'power engineering' is used as a descriptor in that industry. Again, in recent years there has been a growth of new separate-entry degree courses such as 'systems engineering' and 'communication systems engineering', often followed by academic departments of similar name, which are typically not considered as subfields of electronics engineering but of electrical engineering.

Subfields

Electronic engineering has many subfields. Some of the most popular subfields in electronic engineering; although there are engineers who focus exclusively on one subfield, there are also many who focus on a combination of subfields.

Signal processing deals with the analysis and manipulation of signals. Signals can be either analog, in which case the signal varies continuously according to the information, or digital, in which case the signal varies according to a series of discrete values representing the information.

For analog signals, signal processing may involve the amplification and filtering of audio signals for audio equipment or the modulation and demodulation of signals for telecommunications. For digital signals, signal processing may involve the compression, error checking and error detection of digital signals.

Telecommunications engineering deals with the transmission of information across a channel such as a co-axial cable, optical fiber or free space.

Transmissions across free space require information to be encoded in a carrier wave in order to shift the information to a carrier frequency suitable for transmission, this is known as modulation. Popular analog modulation techniques include amplitude modulation and frequency modulation. The choice of modulation affects the cost and performance of a system and these two factors must be balanced carefully by the engineer.

Once the transmission characteristics of a system are determined, telecommunication engineers design the transmitters and receivers needed for such systems. These two are sometimes combined to form a two-way communication device known as a transceiver. A key consideration in the design of transmitters is their power consumption as this is closely related to their signal strength. If the signal strength of a transmitter is insufficient the signal's information will be corrupted by noise.

Electromagnetics is an in-depth study about the signals that are transmitted in a channel (Wired or Wireless). This includes Basics of Electromagnetic waves, Transmission Lines and Waveguides, Antennas, its types and applications with Radio-Frequency (RF) and Microwaves. Its applications are seen widely in other sub-fields like Telecommunication, Control and Instrumentation Engineering.

Control engineering has a wide range of applications from the flight and propulsion systems of commercial airplanes to the cruise control present in many modern cars. It also plays an important role in industrial automation.

Control engineers often utilize feedback when designing control systems. For example, in a car with cruise control, the vehicle's speed is continuously monitored and fed back to the system which adjusts the engine's power output accordingly. Where there is regular feedback, control theory can be used to determine how the system responds to such feedback.

Instrumentation engineering deals with the design of devices to measure physical quantities such as pressure, flow and temperature. These devices are known as instrumentation.

The design of such instrumentation requires a good understanding of physics that often extends beyond electromagnetic theory. For example, radar guns use the Doppler effect to measure the speed of oncoming vehicles. Similarly, thermocouples use the Peltier–Seebeck effect to measure the temperature difference between two points.

Often instrumentation is not used by itself, but instead as the sensors of larger electrical systems. For example, a thermocouple might be used to help ensure a furnace's temperature remains constant. For this reason, instrumentation engineering is often viewed as the counterpart of control engineering.

Computer engineering deals with the design of computers and computer systems. This may involve the design of new computer hardware, the design of PDAs or the use of computers to control an industrial plant. Development of embedded systems—systems made for specific tasks (e.g., mobile phones)—is also included in this field. This field includes the micro controller and its applications. Computer engineers may also work on a system's software. However, the design of complex software systems is often the domain of software engineering, which is usually considered a separate discipline.

VLSI design engineering VLSI stands for *very large scale integration*. It deals with fabrication of ICs and various electronic components.

Signals and Systems

In electrical engineering, the fundamental quantity of representing some information is called a signal. It does not matter what the information is i-e: Analog or digital information. In mathematics, a signal is a function that conveys some information. In fact any quantity measurable through time over space or any higher dimension can be taken as a signal. A signal could be of any dimension and could be of any form.

Analog Signals

A signal could be an analog quantity that means it is defined with respect to the time. It is a continuous signal. These signals are defined over continuous independent variables. They are difficult to analyze, as they carry a huge number of values. They are very much accurate due to a large sample of values. In order to store these signals , you require an infinite memory because it can achieve infinite values on a real line. Analog signals are denoted by sin waves.

Human Voice

Human voice is an example of analog signals. When you speak, the voice that is produced travel through air in the form of pressure waves and thus belongs to a mathematical function, having independent variables of space and time and a value corresponding to air pressure.

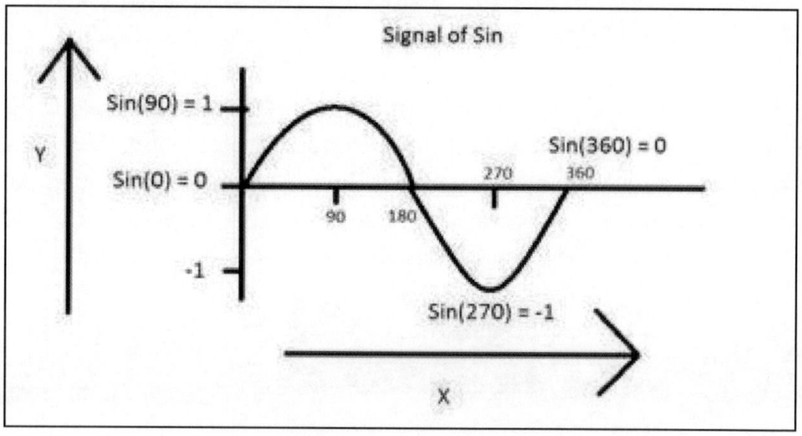

Y = sin(x) where x is independent.

Digital Signals

As compared to analog signals, digital signals are very easy to analyze. They are discontinuous signals. They are the appropriation of analog signals.

The word digital stands for discrete values and hence it means that they use specific values to represent any information. In digital signal, only two values are used to represent something i-e: 1 and 0 (binary values). Digital signals are less accurate then analog signals because they are the discrete samples of an analog signal taken over some period of time. However digital signals are not subject to noise. So they last long and are easy to interpret. Digital signals are denoted by square waves.

Computer Keyboard

Whenever a key is pressed from the keyboard, the appropriate electrical signal is sent to keyboard controller containing the ASCII value that particular key. For example the electrical signal that is generated when keyboard key a is pressed, carry information of digit 97 in the form of 0 and 1, which is the ASCII value of character a.

Difference between Analog and Digital Signals

Comparison element	Analog signal	Digital signal
Analysis	Difficult	Possible to analyze
Representation	Continuous	Discontinuous
Accuracy	More accurate	Less accurate
Storage	Infinite memory	Easily stored
Subject to Noise	Yes	No
Recording Technique	Original signal is preserved	Samples of the signal are taken and preserved
Examples	Human voice, Thermometer, Analog phones e.t.c	Computers, Digital Phones, Digital pens, e.t.c

Systems

A system is a defined by the type of input and output it deals with. Since we are dealing with signals, so in our case, our system would be a mathematical model, a piece of code/software, or a physical device, or a black box whose input is a signal and it performs some processing on that signal, and the output is a signal. The input is known as excitation and the output is known as response.

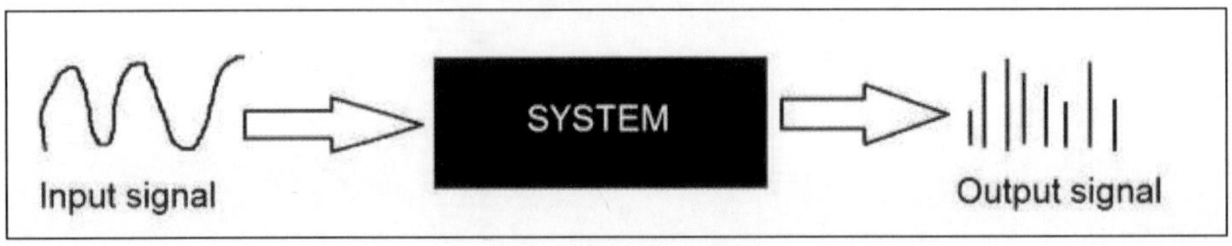

In the above figure a system has been shown whose input and output both are signals but the input is an analog signal. And the output is an digital signal. It means our system is actually a conversion system that converts analog signals to digital signals.

Conversion of Analog to Digital Signals

Since there are lot of concepts related to this analog to digital conversion and vice-versa. We will only discuss those which are related to digital image processing.There are two main concepts that are involved in the coversion:

- Sampling
- Quantization

Sampling

Sampling as its name suggests can be defined as take samples. Take samples of a digital signal over x axis. Sampling is done on an independent variable. In case of this mathematical equation:

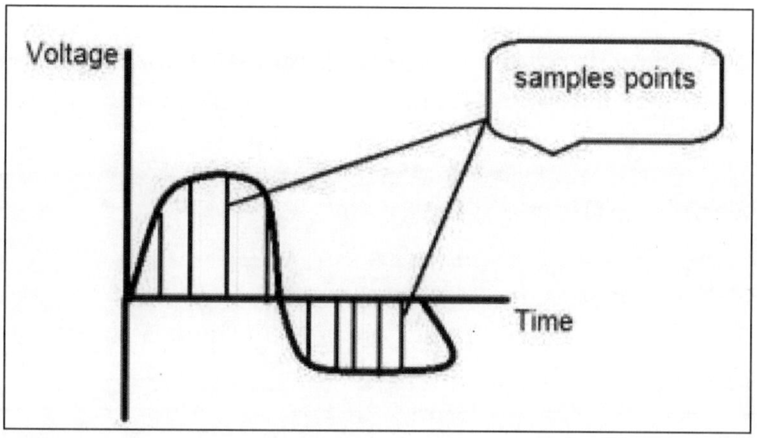

Sampling is done on the x variable. We can also say that the conversion of x axis (infinite values) to digital is done under sampling.

Sampling is further divide into up sampling and down sampling. If the range of values on x-axis are less then we will increase the sample of values. This is known as up sampling and its vice versa is known as down sampling.

Quantization

Quantization as its name suggest can be defined as dividing into quanta (partitions). Quantization is done on dependent variable. It is opposite to sampling.

In case of this mathematical equation y = sin(x). Quantization is done on the Y variable. It is done on the y axis. The conversion of y axis infinite values to 1, 0, -1 (or any other level) is known as Quantization. These are the two basics steps that are involved while converting an analog signal to a digital signal. The quantization of a signal has been shown in the figure.

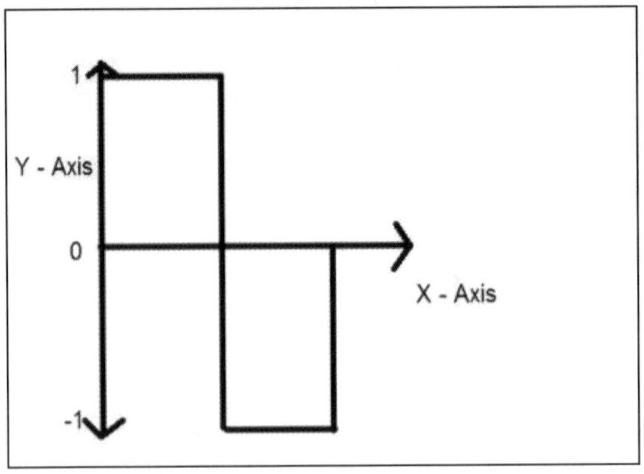

Need to Convert an Analog Signal to Digital Signal

The first and obvious reason is that digital image processing deals with digital images, that are digital signals. So when ever the image is captured, it is converted into digital format and then it is processed.

The second and important reason is, that in order to perform operations on an analog signal with a digital computer, you have to store that analog signal in the computer. And in order to store an analog signal, infinite memory is required to store it. And since that's not possible, so thats why we convert that signal into digital format and then store it in digital computer and then performs operations on it.

Continuous Systems vs. Discrete Systems

Continuous Systems

The type of systems whose input and output both are continuous signals or analog signals are called continuous systems.

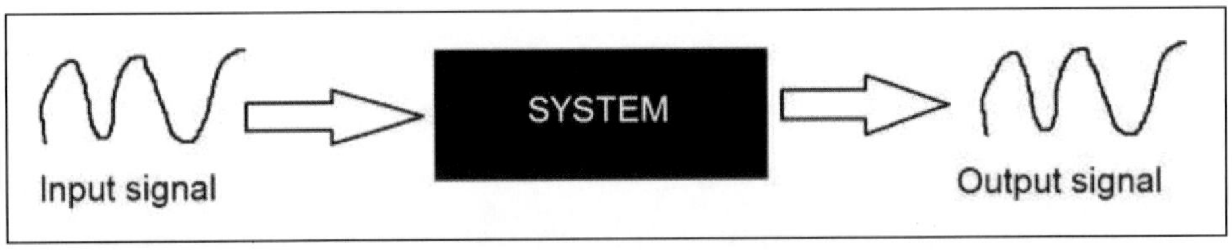

Discrete Systems

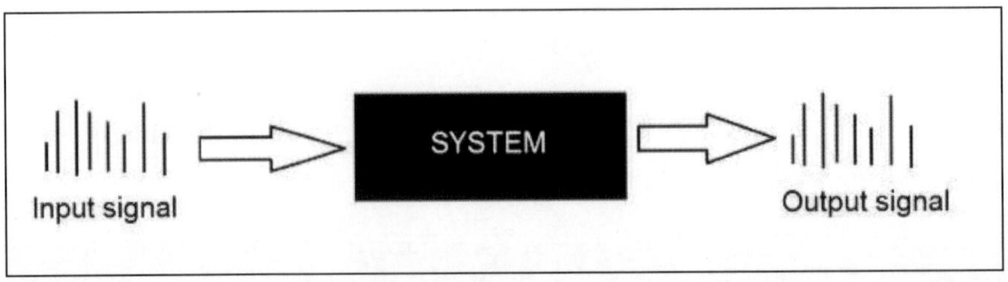

The type of systems whose input and output both are discrete signals or digital signals are called digital systems.

References

- Anthony J. Pansini Electrical Distribution Engineering, p. Xiv, The Fairmont Press Inc., 2006 ISBN 978-0-88173-546-8
- Electronics-and-its-applications, dc-circuits: codrey.com, Retrieved 26 July, 2019
- David K. Ferry/Jonathan P. Bird Electronic Materials and Devices, Academic Press, 2001 ISBN 978-0-12-254161-2
- Signals-and-system-introduction, dip: tutorialspoint.com, Retrieved 25 February, 2019
- Hwei Pia Hsu, Schaum's Outline of Analog and Digital Communications, mcgraw–Hill Professional, 2003. ISBN 978-0-07-140228-6

2
Analog Electronics

The electronic systems with a continuously variable signal are referred to as analog electronics. A few examples of devices which involve the use of analog electronics are voltage amplifiers, variable-gain amplifiers, buffer amplifiers, current amplifiers, power amplifiers, etc. The chapter closely examines these key examples of analog electronics to provide an extensive understanding of the subject.

Analog electronics is a branch of electronics that deals with a continuously variable signal. It's widely used in radio and audio equipment along with other applications where signals are derived from analog sensors before being converted into digital signals for subsequent storage and processing. Although digital circuits are considered as a dominant part of today's technological world, some of the most fundamental components in a digital system are actually analog in nature.

Analog means continuous and real. The world we live in is analog in nature, implying that it's full of infinite possibilities. The number of smells we can sense, the number of tones we can hear, or the number of colors we can paint with; everything is infinite. The people working in the field of analog electronics are basically dealing with analog devices and circuits.

For example, if we build a circuit and it counts values like 1, 2, 3, 4, and 5; the values are neither infinite nor continuous. On the other hand, if the circuit counts like 1.00000, 1.00001, 1.00002, 4.99999, 5.00000, the amount of information would be infinite.

Analog Signals

Before proceeding with analog signals, let's understand the simple meaning of a signal. In electrical engineering, signals are basically time-varying quantities (usually voltage or current). So when we talk about the signal it means we are talking about a voltage that's changing over time.

Signals are passed among devices in order to obtain or send information in the form of audio, video or encoded data. The transmission takes place through wires or via air through radio frequency waves. For instance, audio signals are transferred from the computer's audio card to the speakers, while data signals between a tablet and a Wi-Fi router pass through the air.

Analog Signals use attributes of the medium to convey the signal's information. For example, an aneroid barometer makes use of the angular position of a needle to convey the changes in atmospheric pressure. The signals take any value from a given range and each signal value denotes different information. Each level of the signal signifies a different level of the phenomenon and any change in the signal is meaningful.

It's quite easy to determine if a signal graph is analog or digital. The former one is smooth and continuous while the latter is edge and appears in the form of stepping squares. Given below is an analog signal graph representing the change in voltage with the change in time.

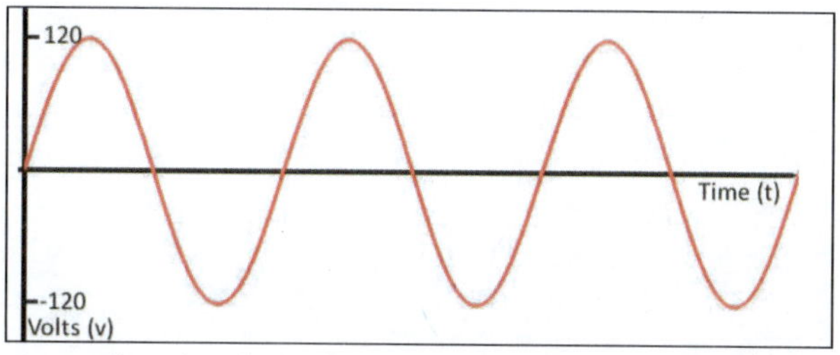

An analog signal graph representing time-voltage graph.

Analog Circuits

Analog circuits can be defined as a complex combination of op amps, resistors, caps, and other basic electronic components. These circuits can be as simple as a combination of two resistors to make a voltage divider or elegantly built with many components. Such circuits can attenuate, amplify, isolate, modify, distort the signal, or even convert the original one into a digital signal.

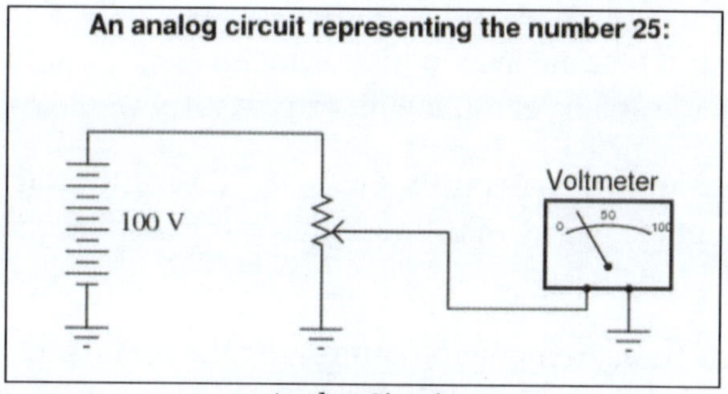

Analog Circuit.

These circuits are difficult to design and need a lot of precision as compared to the digital circuits. The modern circuits are rarely found to be completely analog as these days analog circuitry may use digital or microprocessor techniques to improve performance. Such circuits are called mixed signals.

There are two types of analog circuits namely passive and active, where the former ones don't consume any electrical power while the latter ones do.

Analog vs. Digital Electronics

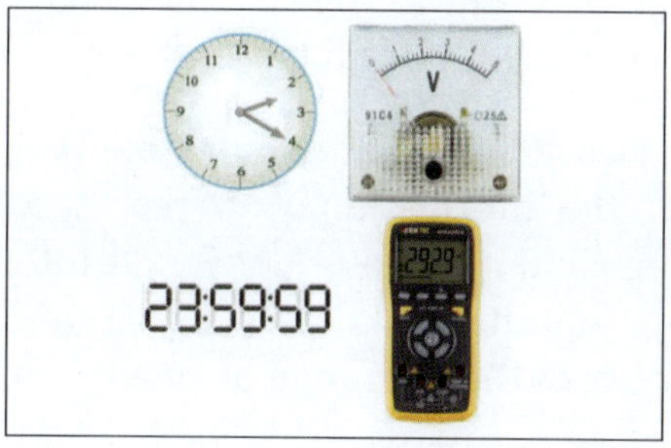

The most basic difference between analog and digital electronics is that in the former technology translates the information into electric pulses of varying amplitude, while the latter translates information into a binary format of 0 and 1, where each bit represents two distinct amplitudes.

However, there are certain characteristic features that clearly state the difference between analog and digital electronics.

Factors	Analog	Digital
Waves	Denoted by Sine waves	Denoted by Square waves
Singal	Continuous signal representing physical Measurments	Discrete signal representing discrete time signals generated by digital modulatin
DataTransmission	Subject to deterioration by noise	Noise-immune without deterioration
Bandwidth	Consumes less bandwidth	Consumes more bandwith
Memory	Stored in the form of wave singnal	Stored in the form of binary bit
Power	Draws large power	Draws negligible power
Impedance	Low impedance	High order of 100 megaohm
Errors	Analog instruments have considerable observational errors	Digital instruments are free from observational errors

Use Analog Components and Circuitry

Analog components are the most basic type of components while the digital components are widely used in today's devices. Both the branches are independently significant in their own way so we are enlisting certain cases when the analog components and circuitry would be ideal.

Filtering Signals: While dealing with a continuous signal, a continuous analog filter is needed to remove all possible frequency content that you don't want. As compared to a digital filter, it is much easier to use as well as costs less.

High Power: Although digital measurement and control may be useful for high power systems, a digital signal switching from 0 to 400V may not be efficient enough. The continuous nature of power delivery in the AC and DC systems require analog components as these are more durable and characterized.

Before A/D and After D/A: In order to switch between continuous and discrete data, hybrid systems need both Analog-to-Digital and Digital-to-Analog converters, but as per the Nyquist's Theorem, the sampling frequency must be at least twice the frequency of the highest frequency component contained in the signal. So, in order to fulfill the theorem, any signal inadvertently included in the original signal can be filtered so as to remove noise after sampling.

Since the signal is not yet digital, the only option is to filter the signal with an analog unit. Once you are done operating on a signal digitally and convert it back to analog, all the processing must once again be done with analog components and circuitry.

Sensors: Sensors convert the real world information into data that can be recognized by a computer or an embedded system. Often the information is not readily available in the real world; rather the sensors first create an analog signal and later convert it into digital signals. Unlike the high voltage systems, sensors have low amplitude and need signal conditioning to increase the value of the signal so as to make a better use of the full range of an ADC.

Analog Signal

An analog signal is any continuous signal for which the time-varying feature (variable) of the signal is a representation of some other time varying quantity, i.e., *analogous* to another time varying signal. For example, in an analog audio signal, the instantaneous voltage of the signal varies continuously with the pressure of the sound waves. It differs from a digital signal, in which the continuous quantity is a representation of a sequence of discrete values which can only take on one of a finite number of values. The term *analog signal* usually refers to electrical signals; however, mechanical, pneumatic, hydraulic, human speech, and other systems may also convey or be considered analog signals.

Representation

An analog signal uses some property of the medium to convey the signal's information. For example, an aneroid barometer uses rotary position as the signal to convey pressure information. In an electrical signal, the voltage, current, or frequency of the signal may be varied to represent the information.

Any information may be conveyed by an analog signal; often such a signal is a measured response to changes in physical phenomena, such as sound, light, temperature, position, or pressure. The physical variable is converted to an analog signal by a transducer. For example, sound striking the diaphragm of a microphone induces corresponding fluctuations in the current produced by a coil in an electromagnetic microphone, or the voltage produced by a condenser microphone. The voltage or the current is said to be an "analog" of the sound.

Noise

An analog signal is subject to electronic noise and distortion introduced by communication channels and signal processing operations, which can progressively degrade the signal-to-noise ratio (SNR). In contrast, digital signals have a finite resolution. Converting an analog signal to digital form introduces a low-level quantization noise into the signal, but once in digital form the signal can be processed or transmitted without introducing significant additional noise or distortion. In analog systems, it is difficult to detect when such degradation occurs. However, in digital systems, degradation can not only be detected but corrected as well.

The most serious disadvantage of analog signals compared to digital transmission is that analog transmissions always contain noise. As the signal is copied, transmitted, or processed, the unavoidable noise introduced in the signal path will accumulate as a generation loss, progressively and irreversibly degrading the signal-to-noise ratio, until in extreme cases the signal can be overwhelmed. Noise can show up as "hiss" and intermodulation distortion in audio signals, or "snow" in video signals. Generation loss is irreversible as there is no reliable method to distinguish the noise from the signal, partly because amplifying the signal to recover attenuated parts of the signal amplifies the noise as well. Digital signals can be transmitted, stored and processed without introducing noise.

In electrical analog signals, noise can be minimized by shielding, good connections and the use of certain cable types such as coaxial or twisted pair.

Amplifiers

A 100 watt stereo audio amplifier used in home component audio systems in the 1970s.

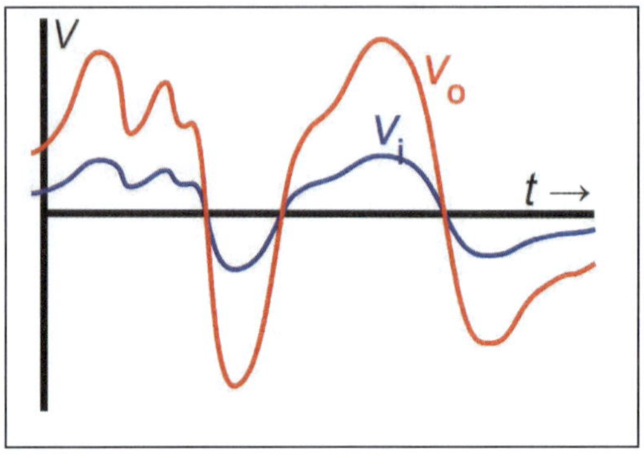

Amplification means increasing the amplitude (voltage or current) of a time-varying signal by a given factor, as shown here. The graph shows the input $v_i(t)$ *(blue)* and output voltage $v_o(t)$ *(red)* of an ideal linear amplifier with an arbitrary signal applied as input. In this example the amplifier has a *voltage gain* of 3; that is at any instant $v_o = 3v_i$.

An amplifier, electronic amplifier or (informally) amp is an electronic device that can increase the power of a signal (a time-varying voltage or current). It is a two-port electronic circuit that uses electric power from a power supply to increase the amplitude of a signal applied to its input terminals, producing a proportionally greater amplitude signal at its output. The amount of amplification provided by an amplifier is measured by its gain: the ratio of output voltage, current, or power to input. An amplifier is a circuit that has a power gain greater than one.

An amplifier can either be a separate piece of equipment or an electrical circuit contained within another device. Amplification is fundamental to modern electronics, and amplifiers are widely used in almost all electronic equipment. Amplifiers can be categorized in different ways. One is by the frequency of the electronic signal being amplified. For example, audio amplifiers amplify signals in the audio (sound) range of less than 20 kHz, RF amplifiers amplify frequencies in the radio frequency range between 20 kHz and 300 GHz, and servo amplifiers and instrumentation amplifiers may work with very low frequencies down to direct current. Amplifiers can also be categorized by their physical placement in the signal chain; a preamplifier may precede other signal processing stages, for example. The first practical electrical device which could amplify was the triode vacuum tube, invented in 1906 by Lee De Forest, which led to the first amplifiers around 1912. Today most amplifiers use transistors.

Ideal

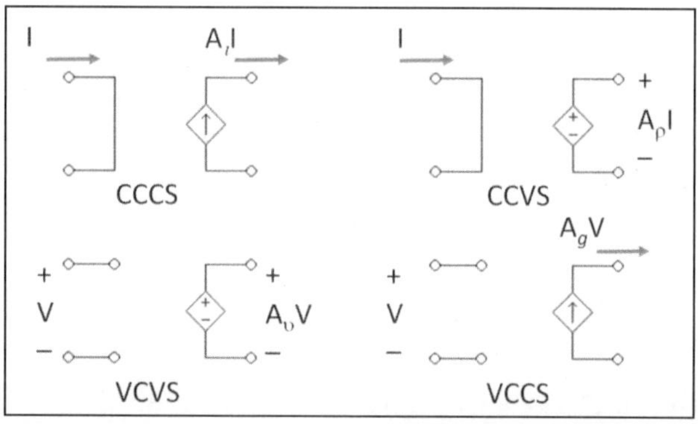

The four types of dependent source—control variable on left, output variable on right.

In principle, an amplifier is an electrical two-port network that produces a signal at the output port that is a replica of the signal applied to the input port, but increased in magnitude.

The input port can be idealized as either being a voltage input, which takes no current, with the output proportional to the voltage across the port; or a current input, with no voltage across it, in which the output is proportional to the current through the port. The output port can be idealized as being either a dependent voltage source, with zero source resistance and its output voltage dependent on the input; or a dependent current source, with infinite source resistance and the output current dependent on the input. Combinations of these choices lead to four types of ideal amplifiers. In idealized form they are represented by each of the four types of dependent source used in linear analysis namely:

Input	Output	Dependent source	Amplifier type	Gain units
I	I	Current controlled current source, CCCS	Current amplifier	Unitless
I	V	Current controlled voltage source, CCVS	Transresistance amplifier	Ohm
V	I	Voltage controlled current source, VCCS	Transconductance amplifier	Siemens
V	V	Voltage controlled voltage source, VCVS	Voltage amplifier	Unitless

Each type of amplifier in its ideal form has an ideal input and output resistance that is the same as that of the corresponding dependent source:

Amplifier type	Dependent source	Input impedance	Output impedance
Current	CCCS	0	∞
Transresistance	CCVS	0	0
Transconductance	VCCS	∞	∞
Voltage	VCVS	∞	0

In real amplifiers the ideal impedances are not possible to achieve, but these ideal elements can be used to construct equivalent circuits of real amplifiers by adding impedances (resistance, capacitance and inductance) to the input and output. For any particular circuit, a small-signal analysis is often used to find the actual impedance. A small-signal AC test current I_x is applied to the input or output node, all external sources are set to AC zero, and the corresponding alternating voltage V_x across the test current source determines the impedance seen at that node as $R = V_x / I_x$.

Amplifiers designed to attach to a transmission line at input and output, especially RF amplifiers, do not fit into this classification approach. Rather than dealing with voltage or current individually, they ideally couple with an input or output impedance matched to the transmission line impedance, that is, match *ratios* of voltage to current. Many real RF amplifiers come close to this ideal. Although, for a given appropriate source and load impedance, RF amplifiers can be characterized as amplifying voltage or current, they fundamentally are amplifying power.

Properties

Amplifier properties are given by parameters that include:

- Gain, the ratio between the magnitude of output and input signals;
- Bandwidth, the width of the useful frequency range;
- Efficiency, the ratio between the power of the output and total power consumption;
- Linearity, the extent to which the proportion between input and output amplitude is the same for high amplitude and low amplitude input;
- Noise, a measure of undesired noise mixed into the output;
- Output dynamic range, the ratio of the largest and the smallest useful output levels;
- Slew rate, the maximum rate of change of the output;
- Rise time, settling time, ringing and overshoot that characterize the step response;
- Stability, the ability to avoid self-oscillation.

Amplifiers are described according to the properties of their inputs, their outputs, and how they relate. All amplifiers have gain, a multiplication factor that relates the magnitude of some property of the output signal to a property of the input signal. The gain may be specified as the ratio of output voltage to input voltage (voltage gain), output power to input power (power gain), or some combination of current, voltage, and power. In many cases the property of the output that varies is dependent on the same property of the input, making the gain unitless (though often expressed in decibels (dB)).

Most amplifiers are designed to be linear. That is, they provide constant gain for any normal input level and output signal. If an amplifier's gain is not linear, the output signal can become distorted. There are, however, cases where variable gain is useful. Certain signal processing applications use exponential gain amplifiers.

Amplifiers are usually designed to function well in a specific application, for example: radio and television transmitters and receivers, high-fidelity ("hi-fi") stereo equipment, microcomputers and other digital equipment, and guitar and other instrument amplifiers. Every amplifier includes at least one active device, such as a vacuum tube or transistor.

Negative Feedback

Negative feedback is a technique used in most modern amplifiers to improve bandwidth and distortion and control gain. In a negative feedback amplifier part of the output is fed back and added to the input in opposite phase, subtracting from the input. The main effect is to reduce the overall gain of the system. However, any unwanted signals introduced by the amplifier, such as distortion are also fed back. Since they are not part of the original input, they are added to the input in opposite phase, subtracting them from the input. In this way, negative feedback also reduces nonlinearity, distortion and other errors introduced by the amplifier. Large amounts of negative feedback can reduce errors to the point that the response of the amplifier itself becomes almost irrelevant as long as it has a large gain, and the output performance of the system (the "closed loop performance") is defined entirely by the components in the feedback loop. This technique is particularly used with operational amplifiers (op-amps).

Non-feedback amplifiers can only achieve about 1% distortion for audio-frequency signals. With negative feedback, distortion can typically be reduced to 0.001%. Noise, even crossover distortion, can be practically eliminated. Negative feedback also compensates for changing temperatures, and degrading or nonlinear components in the gain stage, but any change or nonlinearity in the components in the feedback loop will affect the output. Indeed, the ability of the feedback loop to define the output is used to make active filter circuits.

Another advantage of negative feedback is that it extends the bandwidth of the amplifier. The concept of feedback is used in operational amplifiers to precisely define gain, bandwidth, and other parameters entirely based on the components in the feedback loop.

Negative feedback can be applied at each stage of an amplifier to stabilize the operating point of active devices against minor changes in power-supply voltage or device characteristics.

Some feedback, positive or negative, is unavoidable and often undesirable—introduced, for example, by parasitic elements, such as inherent capacitance between input and output of devices such as transistors, and capacitive coupling of external wiring. Excessive frequency-dependent positive feedback can produce parasitic oscillation and turn an amplifier into an oscillator.

Categories

Active Devices

All amplifiers include some form of active device: this is the device that does the actual amplification. The active device can be a vacuum tube, discrete solid state component, such as a single transistor, or part of an integrated circuit, as in an op-amp).

Transistor amplifiers (or solid state amplifiers) are the most common type of amplifier in use today. A transistor is used as the active element. The gain of the amplifier is determined by the properties of the transistor itself as well as the circuit it is contained within.

Common active devices in transistor amplifiers include bipolar junction transistors (BJTs) and metal oxide semiconductor field-effect transistors (MOSFETs). Applications are numerous, some common examples are audio amplifiers in a home stereo or public address system, RF high power generation for semiconductor equipment, to RF and microwave applications such as radio transmitters.

Transistor-based amplification can be realized using various configurations: for example a bipolar junction transistor can realize common base, common collector or common emitter amplification;

a MOSFET can realize common gate, common source or common drain amplification. Each configuration has different characteristics.

Vacuum-tube amplifiers (also known as tube amplifiers or valve amplifiers) use a vacuum tube as the active device. While semiconductor amplifiers have largely displaced valve amplifiers for low-power applications, valve amplifiers can be much more cost effective in high power applications such as radar, countermeasures equipment, and communications equipment. Many microwave amplifiers are specially designed valve amplifiers, such as the klystron, gyrotron, traveling wave tube, and crossed-field amplifier, and these microwave valves provide much greater single-device power output at microwave frequencies than solid-state devices. Vacuum tubes remain in use in some high end audio equipment, as well as in musical instrument amplifiers, due to a preference for "tube sound".

Magnetic amplifiers are devices somewhat similar to a transformer where one winding is used to control the saturation of a magnetic core and hence alter the impedance of the other winding.

They have largely fallen out of use due to development in semiconductor amplifiers but are still useful in HVDC control, and in nuclear power control circuitry due to not being affected by radioactivity. Negative resistances can be used as amplifiers, such as the tunnel diode amplifier.

Power Amplifiers

A power amplifier is an amplifier designed primarily to increase the power available to a load. In practice, amplifier power gain depends on the source and load impedances, as well as the inherent voltage and current gain. A radio frequency (RF) amplifier design typically optimizes impedances for power transfer, while audio and instrumentation amplifier designs normally optimize input and output impedance for least loading and highest signal integrity. An amplifier that is said to have a gain of 20 dB might have a voltage gain of 20 dB and an available power gain of much more than 20 dB (power ratio of 100)—yet actually deliver a much lower power gain if, for example, the input is from a 600 Ω microphone and the output connects to a 47 kΩ input socket for a power amplifier. In general the power amplifier is the last 'amplifier' or actual circuit in a signal chain (the output stage) and is the amplifier stage that requires attention to power efficiency. Efficiency considerations lead to the various classes of power amplifier based on the biasing of the output transistors or tubes.

Power amplifier by Skyworks Solutions in a Smartphone.

Audio power amplifiers are typically used to drive loudspeakers. They will often have two output channels and deliver equal power to each. An RF power amplifier is found in radio transmitter final stages. A Servo motor controller: amplifies a control voltage to adjust the speed of a motor, or the position of a motorized system.

Operational Amplifiers (Op-amps)

An LM741 general purpose op-amp.

An operational amplifier is an amplifier circuit which typically has very high open loop gain and differential inputs. Op amps have become very widely used as standardized "gain blocks" in circuits due to their versatility; their gain, bandwidth and other characteristics can be controlled by feedback through an external circuit. Though the term today commonly applies to integrated circuits, the original operational amplifier design used valves, and later designs used discrete transistor circuits.

A fully differential amplifier is similar to the operational amplifier, but also has differential outputs. These are usually constructed using BJTs or FETs.

Distributed Amplifiers

These use balanced transmission lines to separate individual single stage amplifiers, the outputs of which are summed by the same transmission line. The transmission line is a balanced type with the input at one end and on one side only of the balanced transmission line and the output at the opposite end is also the opposite side of the balanced transmission line. The gain of each stage adds linearly to the output rather than multiplies one on the other as in a cascade configuration. This allows a higher bandwidth to be achieved than could otherwise be realised even with the same gain stage elements.

Switched Mode Amplifiers

These nonlinear amplifiers have much higher efficiencies than linear amps, and are used where the power saving justifies the extra complexity. Class-D amplifiers are the main example of this type of amplification.

Negative Resistance Amplifier

Negative Resistance Amplifier is a type of Regenerative Amplifier that can use the feedback between the transistor's source and gate to transform a capacitive impedance on the transistor's source to a negative resistance on its gate. Compared to other types of amplifiers, this "negative resistance amplifier" will only require a tiny amount of power to achieve very high gain, maintaining a good noise figure at the same time.

Applications

Video Amplifiers

Video amplifiers are designed to process video signals and have varying bandwidths depending on whether the video signal is for SDTV, EDTV, HDTV 720p or 1080i/p etc.. The specification of the bandwidth itself depends on what kind of filter is used—and at which point (−1 dB or −3 dB for example) the bandwidth is measured. Certain requirements for step response and overshoot are necessary for an acceptable TV image.

Microwave Amplifiers

Traveling wave tube amplifiers (TWTAs) are used for high power amplification at low microwave frequencies. They typically can amplify across a broad spectrum of frequencies; however, they are usually not as tunable as klystrons.

Klystrons are specialized linear-beam vacuum-devices, designed to provide high power, widely tunable amplification of millimetre and sub-millimetre waves. Klystrons are designed for large scale operations and despite having a narrower bandwidth than TWTAs, they have the advantage of coherently amplifying a reference signal so its output may be precisely controlled in amplitude, frequency and phase.

Solid-state devices are used such as GaAs FETs, IMPATT diodes, and others, especially at lower microwave frequencies and power levels on the order of watts. The maser is a non-electronic microwave amplifier.

Musical Instrument Amplifiers

Instrument amplifiers are a range of audio power amplifiers used to increase the sound level of musical instruments, for example guitars, during performances.

Classification of Amplifier Stages and Systems

Common Terminal

One set of classifications for amplifiers is based on which device terminal is common to both the input and the output circuit. In the case of bipolar junction transistors, the three classes are common emitter, common base, and common collector. For field-effect transistors, the corresponding configurations are common source, common gate, and common drain; for vacuum tubes, common cathode, common grid, and common plate.

The common emitter (or common source, common cathode, etc.) is most often configured to provide amplification of a voltage applied between base and emitter, and the output signal taken between collector and emitter is inverted, relative to the input. The common collector arrangement applies the input voltage between base and collector, and to take the output voltage between emitter and collector. This causes negative feedback, and the output voltage tends to follow the input voltage. This arrangement is also used as the input presents a high impedance and does not load the signal source, though the voltage amplification is less than one. The common-collector circuit is, therefore, better known as an emitter follower, source follower, or cathode follower.

Unilateral or Bilateral

An amplifier whose output exhibits no feedback to its input side is described as 'unilateral'. The input impedance of a unilateral amplifier is independent of load, and output impedance is independent of signal source impedance.

An amplifier that uses feedback to connect part of the output back to the input is a *bilateral* amplifier. Bilateral amplifier input impedance depends on the load, and output impedance on the signal source impedance. All amplifiers are bilateral to some degree; however they may often be modeled as unilateral under operating conditions where feedback is small enough to neglect for most purposes, simplifying analysis.

Inverting or Non-inverting

Another way to classify amplifiers is by the phase relationship of the input signal to the output signal. An 'inverting' amplifier produces an output 180 degrees out of phase with the input signal (that is, a polarity inversion or mirror image of the input as seen on an oscilloscope). A 'non-inverting' amplifier maintains the phase of the input signal waveforms. An emitter follower is a type of non-inverting amplifier, indicating that the signal at the emitter of a transistor is following (that is, matching with unity gain but perhaps an offset) the input signal. Voltage follower is also non inverting type of amplifier having unity gain.

This description can apply to a single stage of an amplifier, or to a complete amplifier system.

Function

Other amplifiers may be classified by their function or output characteristics. These functional descriptions usually apply to complete amplifier systems or sub-systems and rarely to individual stages.

- A servo amplifier indicates an integrated feedback loop to actively control the output at some desired level. A DC servo indicates use at frequencies down to DC levels, where the rapid fluctuations of an audio or RF signal do not occur. These are often used in mechanical actuators, or devices such as DC motors that must maintain a constant speed or torque. An AC servo amp. can do this for some AC motors.

- A linear amplifier responds to different frequency components independently, and does not generate harmonic distortion or intermodulation distortion. No amplifier can provide *perfect* linearity (even the most linear amplifier has some nonlinearities, since the amplifying devices—transistors or vacuum tubes—follow nonlinear power laws such as square-laws and rely on circuitry techniques to reduce those effects).

- A nonlinear amplifier generates significant distortion and so changes the harmonic content; there are situations where this is useful. Amplifier circuits intentionally providing a non-linear transfer function include:
    - A device like a silicon controlled rectifier or a transistor used as a switch may be employed to turn either fully on or off a load such as a lamp based on a threshold in a continuously variable input.
    - A non-linear amplifier in an analog computer or true RMS converter for example can provide a special transfer function, such as logarithmic or square-law.
    - A Class C RF amplifier may be chosen because it can be very efficient—but is non-linear. Following such an amplifier with a so-called tank tuned circuit can reduce unwanted harmonics (distortion) sufficiently to make it useful in transmitters, or some desired harmonic may be selected by setting the resonant frequency of the tuned circuit to a higher frequency rather than fundamental frequency in frequency multiplier circuits.
    - Automatic gain control circuits require an amplifier's gain be controlled by the time-averaged amplitude so that the output amplitude varies little when weak stations are being received. The non-linearities are assumed arranged so the relatively small signal amplitude suffers from little distortion (cross-channel interference or intermodulation) yet is still modulated by the relatively large gain-control DC voltage.
    - AM detector circuits that use amplification such as anode-bend detectors, precision rectifiers and infinite impedance detectors (so excluding unamplified detectors such as cat's-whisker detectors), as well as peak detector circuits, rely on changes in amplification based on the signal's instantaneous amplitude to derive a direct current from an alternating current input.
    - Operational amplifier comparator and detector circuits.
- A wideband amplifier has a precise amplification factor over a wide frequency range, and is often used to boost signals for relay in communications systems. A narrowband amp amplifies a specific narrow range of frequencies, to the exclusion of other frequencies.
- An RF amplifier amplifies signals in the radio frequency range of the electromagnetic spectrum, and is often used to increase the sensitivity of a receiver or the output power of a transmitter.
- An audio amplifier amplifies audio frequencies. This category subdivides into small signal amplification, and power amps that are optimised to driving speakers, sometimes with multiple amps grouped together as separate or bridgeable channels to accommodate different audio reproduction requirements. Frequently used terms within audio amplifiers include:
    - Preamplifier (preamp.), which may include a phono preamp with RIAA equalization, or tape head preamps with CCIR equalisation filters. They may include filters or tone control circuitry.

- Power amplifier (normally drives loudspeakers), headphone amplifiers, and public address amplifiers.
- Stereo amplifiers imply two channels of output (left and right), though the term simply means "solid" sound (referring to three-dimensional)—so quadraphonic stereo was used for amplifiers with four channels. 5.1 and 7.1 systems refer to Home theatre systems with 5 or 7 normal spatial channels, plus a subwoofer channel.

- Buffer amplifiers, which may include emitter followers, provide a high impedance input for a device (perhaps another amplifier, or perhaps an energy-hungry load such as lights) that would otherwise draw too much current from the source. Line drivers are a type of buffer that feeds long or interference-prone interconnect cables, possibly with differential outputs through twisted pair cables.

Interstage Coupling Method

Amplifiers are sometimes classified by the coupling method of the signal at the input, output, or between stages. Different types of these include:

Resistive-capacitive (RC) coupled amplifier, using a network of resistors and capacitors: By design these amplifiers cannot amplify DC signals as the capacitors block the DC component of the input signal. RC-coupled amplifiers were used very often in circuits with vacuum tubes or discrete transistors. In the days of the integrated circuit a few more transistors on a chip are much cheaper and smaller than a capacitor.

Inductive-capacitive (LC) coupled amplifier, using a network of inductors and capacitors: This kind of amplifier is most often used in selective radio-frequency circuits.

Transformer coupled amplifier, using a transformer to match impedances or to decouple parts of the circuits: Quite often LC-coupled and transformer-coupled amplifiers cannot be distinguished as a transformer is some kind of inductor.

Direct coupled amplifier, using no impedance and bias matching components: This class of amplifier was very uncommon in the vacuum tube days when the anode (output) voltage was at greater than several hundred volts and the grid (input) voltage at a few volts minus. So they were only used if the gain was specified down to DC (e.g., in an oscilloscope). In the context of modern electronics developers are encouraged to use directly coupled amplifiers whenever possible. In FET and CMOS technologies direct coupling is dominant since gates of MOSFETs theoretically pass no current through themselves. Therefore, DC component of the input signals is automatically filtered.

Frequency Range

Depending on the frequency range and other properties amplifiers are designed according to different principles.

Frequency ranges down to DC are only used when this property is needed. Amplifiers for direct current signals are vulnerable to minor variations in the properties of components with time. Special methods, such as chopper stabilized amplifiers are used to prevent objectionable drift in the amplifier's properties for DC. "DC-blocking" capacitors can be added to remove DC and sub-sonic frequencies from audio amplifiers.

Depending on the frequency range specified different design principles must be used. Up to the MHz range only "discrete" properties need be considered; e.g., a terminal has an input impedance.

As soon as any connection within the circuit gets longer than perhaps 1% of the wavelength of the highest specified frequency (e.g., at 100 MHz the wavelength is 3 m, so the critical connection length is approx. 3 cm) design properties radically change. For example, a specified length and width of a PCB trace can be used as a selective or impedance-matching entity. Above a few hundred MHz, it gets difficult to use discrete elements, especially inductors. In most cases, PCB traces of very closely defined shapes are used instead (stripline techniques).

The frequency range handled by an amplifier might be specified in terms of bandwidth (normally implying a response that is 3 dB down when the frequency reaches the specified bandwidth), or by specifying a frequency response that is within a certain number of decibels between a lower and an upper frequency (e.g. "20 Hz to 20 kHz plus or minus 1 dB").

Power Amplifier Classes

Power amplifier circuits (output stages) are classified as A, B, AB and C for analog designs—and class D and E for switching designs. The power amplifier classes are based on the proportion of each input cycle (conduction angle) during which an amplifying device passes current. The image of the conduction angle derives from amplifying a sinusoidal signal. If the device is always on, the conducting angle is 360°. If it is on for only half of each cycle, the angle is 180°. The angle of flow is closely related to the amplifier power efficiency.

Amplifier Circuit

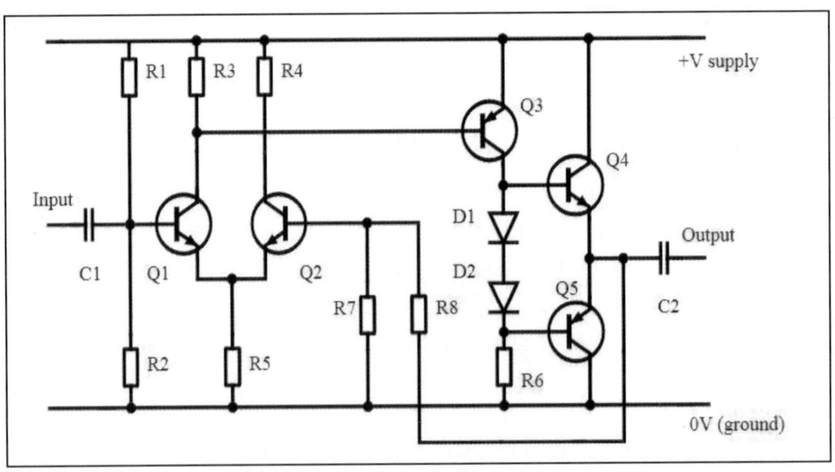

A practical amplifier circuit.

The practical amplifier circuit shown above could be the basis for a moderate-power audio amplifier. It features a typical (though substantially simplified) design as found in modern amplifiers, with a class-AB push–pull output stage, and uses some overall negative feedback. Bipolar transistors are shown, but this design would also be realizable with FETs or valves.

The input signal is coupled through capacitor C1 to the base of transistor Q1. The capacitor allows the AC signal to pass, but blocks the DC bias voltage established by resistors R1 and R2 so that any preceding circuit is not affected by it. Q1 and Q2 form a differential amplifier (an amplifier that multiplies the difference between two inputs by some constant), in an arrangement known as a long-tailed pair. This arrangement is used to conveniently allow the use of negative feedback, which is fed from the output to Q2 via R7 and R8.

The negative feedback into the difference amplifier allows the amplifier to compare the input to the actual output. The amplified signal from Q1 is directly fed to the second stage, Q3, which is a common emitter stage that provides further amplification of the signal and the DC bias for the output stages, Q4 and Q5. R6 provides the load for Q3 (a better design would probably use some form of active load here, such as a constant-current sink). So far, all of the amplifier is operating in class A. The output pair are arranged in class-AB push–pull, also called a complementary pair. They provide the majority of the current amplification (while consuming low quiescent current) and directly drive the load, connected via DC-blocking capacitor C2. The diodes D1 and D2 provide a small amount of constant voltage bias for the output pair, just biasing them into the conducting state so that crossover distortion is minimized. That is, the diodes push the output stage firmly into class-AB mode (assuming that the base-emitter drop of the output transistors is reduced by heat dissipation).

This design is simple, but a good basis for a practical design because it automatically stabilises its operating point, since feedback internally operates from DC up through the audio range and beyond. Further circuit elements would probably be found in a real design that would roll-off the frequency response above the needed range to prevent the possibility of unwanted oscillation. Also, the use of fixed diode bias as shown here can cause problems if the diodes are not both electrically and thermally matched to the output transistors – if the output transistors turn on too much, they can easily overheat and destroy themselves, as the full current from the power supply is not limited at this stage.

A common solution to help stabilise the output devices is to include some emitter resistors, typically one ohm or so. Calculating the values of the circuit's resistors and capacitors is done based on the components employed and the intended use of the amp.

Implementation

Any real amplifier is an imperfect realization of an ideal amplifier. An important limitation of a real amplifier is that the output it generates is ultimately limited by the power available from the power supply. An amplifier saturates and clips the output if the input signal becomes too large for the amplifier to reproduce or exceeds operational limits for the device. The power supply may influence the output, so must be considered in the design. The power output from an amplifier cannot exceed its input power.

The amplifier circuit has an "open loop" performance. This is described by various parameters (gain, slew rate, output impedance, distortion, bandwidth, signal-to-noise ratio, etc.). Many modern amplifiers use negative feedback techniques to hold the gain at the desired value and reduce distortion. Negative loop feedback has the intended effect of lowering the output impedance and thereby increasing electrical damping of loudspeaker motion at and near the resonance frequency of the speaker.

When assessing rated amplifier power output, it is useful to consider the applied load, the signal type (e.g., speech or music), required power output duration (i.e., short-time or continuous), and required dynamic range (e.g., recorded or live audio). In high-powered audio applications that require long cables to the load (e.g., cinemas and shopping centres) it may be more efficient to connect to the load at line output voltage, with matching transformers at source and loads. This avoids long runs of heavy speaker cables.

To prevent instability or overheating requires care to ensure solid state amplifiers are adequately loaded. Most have a rated minimum load impedance.

All amplifiers generate heat through electrical losses. The amplifier must dissipate this heat via convection or forced air cooling. Heat can damage or reduce electronic component service life. Designers and installers must also consider heating effects on adjacent equipment.

Different power supply types result in many different methods of bias. Bias is a technique by which active devices are set to operate in a particular region, or by which the DC component of the output signal is set to the midpoint between the maximum voltages available from the power supply. Most amplifiers use several devices at each stage; they are typically matched in specifications except for polarity. Matched inverted polarity devices are called complementary pairs. Class-A amplifiers generally use only one device, unless the power supply is set to provide both positive and negative voltages, in which case a dual device symmetrical design may be used. Class-C amplifiers, by definition, use a single polarity supply.

Amplifiers often have multiple stages in cascade to increase gain. Each stage of these designs may be a different type of amp to suit the needs of that stage. For instance, the first stage might be a class-A stage, feeding a class-AB push–pull second stage, which then drives a class-G final output stage, taking advantage of the strengths of each type, while minimizing their weaknesses.

Voltage Amplifier

A Voltage Amplifier is an electronic circuit whose function is to accept an input voltage and produce a magnified, accurate replica of this voltage as an output voltage. The voltage gain of the amplifier is the amplitude ratio of the output voltage to the input voltage. Often, electronic amplifiers designed to operate in different environments are categorized by criteria other than their voltage gain, even though they are voltage amplifiers in fact. Many specialized circuits are designed to provide voltage amplification.

Voltage amplifiers are distinguished from other categories of amplifiers whose ability to amplify voltages, or lack thereof, is of secondary importance. Amplifiers in other categories usually are designed to deliver power gain (power amplifiers, including push-pull amplifiers) or to isolate one part of a circuit from another (buffers and emitter followers). Power amplifiers may or may not have voltage gain, while buffers and emitter followers generally produce power gain without a corresponding voltage gain.

Transistor amplifiers, such as the junction field-effect transistor (JFET) or the bipolar junction transistor (BJT) amplifier, will not operate properly without proper gate (JFET) or base (BJT) bias voltages applied in series with the signal voltage. These bias circuits can be modeled as ideal voltage sources. The bias and signal voltages are chosen so that the total input voltage—bias plus signal—will not cut off or saturate the amplifier for any value in the range of the input signal voltage. In addition to a bias voltage source, well-designed bipolar transistor amplifiers require negative feedback at dc to protect the transistor from thermal runaway.

To obtain high gain, cascades of single amplifier circuits are used, usually with a coupling network, actually a simple filter, inserted between the stages of amplification. One such filter is a high-pass network formed by a coupling capacitor, the output resistances of the driving stage, and the input resistance of the driven stage. Since dc voltages are blocked by the capacitor, this ac coupling permits independently setting dc bias voltages for each amplifier stage in the cascade. The coupling network also rejects signals with ac frequency components below a cutoff. The capacitor must be sufficiently large not to attenuate any of the frequencies that are to be amplified. If dc is to be amplified, a direct-coupled amplifier is required, and the design is somewhat more complicated since dc bias voltages on each transistor now cannot be set independently.

The amplifiers discussed above are called single-ended amplifiers, since their input and output voltages are referred to a common reference point which by convention is called ground. These single-ended circuits, while satisfactory for most noncritical applications, have several weaknesses which degrade their performance in high-gain, weak-signal applications. Their unbalanced construction and their use of a common ground point for return currents makes them susceptible to noise pickup.

To minimize noise on sensitive signal lines, special balanced differential amplifier circuits are often used in critical amplifier applications. Differential amplifiers are designed to have equal impedances to ground for each side of the signal line and to have an output voltage proportional to the difference of the voltages from each signal line to ground. This symmetry cancels common-mode noise voltages, voltages which tend to appear on each of the signal lines as equal voltages to ground. Proper circuit design, with attention to the symmetry of the input circuit construction, can ensure that the majority of undesired noise pickup will be common-mode noise and, hence, will be attenuated by the differential amplifier.

In cases where a voltage amplifier is required for some special purpose, operational amplifiers are often used to fill the need. The operational amplifier is an integrated circuit containing a cascade of differential amplifier stages, usually followed by a push-pull amplifier acting as a buffer. The differential voltage gain of the operational amplifier is very high, about 100,000 at low frequencies, while its input impedance is in the megohm range and its output impedance is usually under 100 ohms. The amplifier is designed to be used in a negative-feedback configuration, where the desired gain is controlled by a resistive voltage divider feeding a fraction of the output voltage to the inverting input of the operational amplifier.

With needed amplification built into many integrated circuits and with the availability of operational amplifiers for special-purpose amplification needs, there is seldom a need to design and build a voltage amplifier from discrete components.

Variable-gain Amplifier

A variable-gain or voltage-controlled amplifier is an electronic amplifier that varies its gain depending on a control voltage (often abbreviated CV).

VCAs have many applications, including audio level compression, synthesizers and amplitude modulation.

A crude example is a typical inverting op-amp configuration with a light-dependent resistor (LDR) in the feedback loop. The gain of the amplifier then depends on the light falling on the LDR, which can be provided by an LED (an optocoupler). The gain of the amplifier is then controllable by the current through the LED. This is similar to the circuits used in optical audio compressors.

A voltage-controlled amplifier can be realised by first creating a voltage-controlled resistor (VCR), which is used to set the amplifier gain. The VCR is one of the numerous interesting circuit elements that can be produced by using a JFET (junction field-effect transistor) with simple biasing. VCRs manufactured in this way can be obtained as discrete devices, e.g. VCR2N.

Another type of circuit uses operational transconductance amplifiers. In audio applications logarithmic gain control is used to emulate how the ear hears loudness. David E. Blackmer's dbx 202 VCA was among the first successful implementations of a logarithmic VCA.

Sound Mixing Consoles

Some mixing consoles come equipped with VCAs in each channel for console automation. The fader, which traditionally controls the audio signal directly, becomes a DC control voltage for the VCA. The maximum voltage available to a fader can be controlled by one or more master faders called VCA groups. The VCA master fader then controls the overall level of all of the channels assigned to it. Typically VCA groups are used to control various parts of the mix; vocals, guitars, drums or percussion. The VCA master fader allows a portion of a mix to be raised or lowered without affecting the blend of the instruments in that part of the mix.

A benefit of VCA sub-group is that since it is directly affecting the gain level of each channel, changes to a VCA sub-group level affect not only the channel level, but also all of the levels sent to any post fader mixes. With traditional audio sub-groups, the sub-group master fader only affects the level going into the main mix and does not affect the level going to the post fader mixes. Consider the case of an instrument feeding a sub-group and a post fader mix. If you completely lower the sub-group master fader, you would no longer hear the instrument itself, but you would still hear it as part of the post fader mix, perhaps a reverb or chorus effect.

VCA mixers are known to last longer than non- VCA mixers. Because the VCA controls the audio level instead of the physical fader, decay of the fader mechanism over time does not cause a degradation in audio quality.

Digital Variable-gain Amplifier

A digitally controlled amplifier (DCA) is a variable-gain amplifier that is digitally controlled.

The digitally controlled amplifier uses a stepped approach giving the circuit graduated increments of gain selection. This can be done in several fashions, but certain elements remain in any design.

At its most basic form, a toggle switch strapped across the feedback resistor can provide two discrete gain settings. While this is not a computer controlled function, it describes the core function. With eight switches and eight resistors in the feedback loop, each switch can enable a particular resistor to control the feedback of the amplifier. If each switch was converted to a relay, a microcontroller could be used to activate the relays to attain the desired amount of gain.

Relays can be replaced with Field Effect Transistors of an appropriate type to reduce the mechanical nature of the design. Other devices such as the CD4053 bi-directional CMOS analog multiplexer integrated circuit and digital potentiometers (combined resistor string and MUXes) can serve well as the switching function.

To minimize the quantity of switches and resistors, combinations of resistance values can be utilized by activating multiple switches.

Buffer Amplifier

A buffer amplifier (sometimes simply called a buffer) is one that provides electrical impedance transformation from one circuit to another, with the aim of preventing the signal source from being affected by whatever currents (or voltages, for a current buffer) that the load may be produced with. The signal is 'buffered from' load currents. Two main types of buffer exist: the voltage buffer and the current buffer.

Voltage Buffer

A voltage buffer amplifier is used to transfer a voltage from a first circuit, having a high output impedance level, to a second circuit with a low input impedance level. The interposed buffer amplifier prevents the second circuit from loading the first circuit unacceptably and interfering with its desired operation. In the ideal voltage buffer in the diagram, the input resistance is infinite and the output resistance zero (output impedance of an ideal voltage source is zero). Other properties of the ideal buffer are: perfect linearity, regardless of signal amplitudes; and instant output response, regardless of the speed of the input signal.

If the voltage is transferred unchanged (the voltage gain A_v is 1), the amplifier is a unity gain buffer; also known as a voltage follower because the output voltage *follows* or tracks the input voltage. Although the voltage gain of a voltage buffer amplifier may be (approximately) unity, it usually provides considerable current gain and thus power gain. However, it is commonplace to say that it has a gain of 1 (or the equivalent 0 dB), referring to the voltage gain.

As an example, consider a Thévenin source (voltage V_A, series resistance R_A) driving a resistor load R_L. Because of voltage division (also referred to as "loading") the voltage across the load is only $V_A R_L / (R_L + R_A)$. However, if the Thévenin source drives a unity gain buffer such as that in figure (top, with unity gain), the voltage input to the amplifier is V_A, and with *no voltage division* because the amplifier input resistance is infinite. At the output the dependent voltage source delivers voltage $A_v V_A = V_A$ to the load, again without voltage division because the output resistance of the buffer is zero. A Thévenin equivalent circuit of the combined original Thévenin source *and* the buffer is an ideal voltage source V_A with zero Thévenin resistance.

Current Buffer

Typically a current buffer amplifier is used to transfer a current from a first circuit, having a low output impedance level, to a second circuit with a high input impedance level. The interposed buffer amplifier prevents the second circuit from loading the first circuit's current unacceptably and interfering with its desired operation. In the ideal current buffer in the diagram, the output impedance is infinite (an ideal current source) and the input impedance is zero (a short circuit). Again, other properties of the ideal buffer are: perfect linearity, regardless of signal amplitudes; and instant output response, regardless of the speed of the input signal.

For a current buffer, if the current is transferred unchanged (the current gain β_i is 1), the amplifier is again a unity gain buffer; this time known as a current follower because the output current *follows* or tracks the input current.

As an example, consider a Norton source (current I_A, parallel resistance R_A) driving a resistor load R_L. Because of current division (also referred to as "loading") the current delivered to the load is only $I_A R_A / (R_L + R_A)$. However, if the Norton source drives a unity gain buffer such as that in figure (bottom, with unity gain), the current input to the amplifier is I_A, with *no current division* because the amplifier input resistance is zero. At the output the dependent current source delivers current $\beta_i I_A = I_A$ to the load, again without current division because the output resistance of the buffer is infinite. A Norton equivalent circuit of the combined original Norton source *and* the buffer is an ideal current source I_A with infinite Norton resistance.

Voltage Buffer Examples

Op-amp Implementation

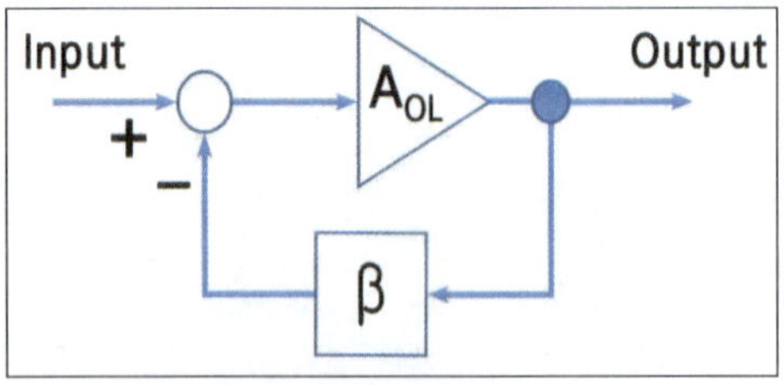

A negative feedback amplifier.

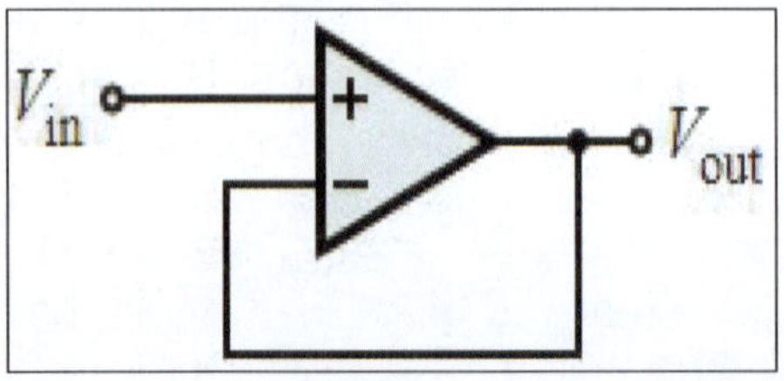

An op-amp–based unity gain buffer amplifier.

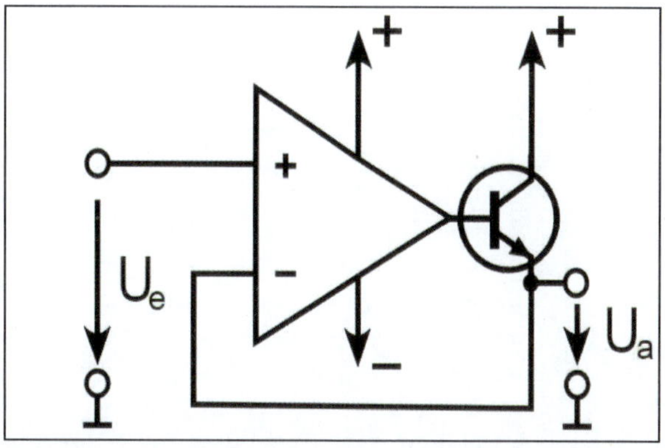

A voltage follower boosted by a transistor; also can be seen as the "ideal transistor" without a base voltage drop on the input signal. This is the basic circuit of linear voltage regulators.

A unity gain buffer amplifier may be constructed by applying a full series negative feedback to an op-amp simply by connecting its output to its inverting input, and connecting the signal source to the non-inverting input. Unity gain here implies a *voltage gain* of one (i.e. 0 dB), but significant *current gain* is expected. In this configuration, the entire output voltage (β = 1) is fed back into the inverting input. The difference between the non-inverting input voltage and the inverting input voltage is amplified by the op-amp. This connection forces the op-amp to adjust its output voltage simply equal to the input voltage (V_{out} follows V_{in} so the circuit is named op-amp voltage follower).

The importance of this circuit does not come from any change in voltage, but from the input and output impedances of the op-amp. The input impedance of the op-amp is very high (1 MΩ to 10 TΩ), meaning that the input of the op-amp does not load down the source and draws only minimal current from it. Because the output impedance of the op-amp is very low, it drives the load as if it were a perfect voltage source. Both the connections to and from the buffer are therefore bridging connections, which reduce power consumption in the source, distortion from overloading, cross-talk and other electromagnetic interference.

Single-transistor Circuits

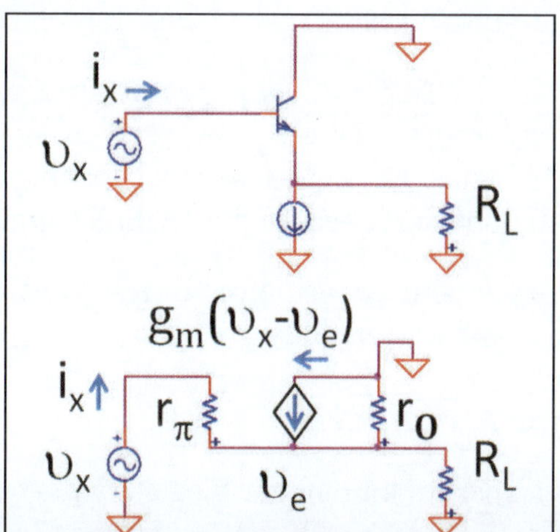

Top: BJT voltage follower, Bottom: Small-signal, low-frequency equivalent circuit using hybrid-pi model.

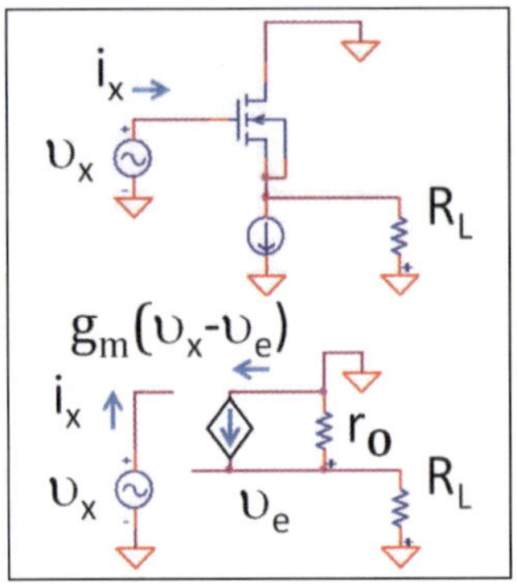

Top: MOSFET voltage follower, Bottom: Small-signal, low-frequency equivalent circuit using hybrid-pi model.

Other unity gain buffer amplifiers include the bipolar junction transistor in common-collector configuration (called an *emitter follower* because the emitter voltage follows the base voltage, or a *voltage follower* because the output voltage follows the input voltage); the field effect transistor in common-drain configuration (called a source follower because the source voltage follows the gate voltage or, again, a *voltage follower* because the output voltage follows the input voltage); or similar configurations using vacuum tubes (cathode follower), or other active devices. All such amplifiers actually have a gain of slightly less than unity, but the difference is usually small and unimportant.

Impedance Transformation using the Bipolar Voltage Follower

Using the small-signal circuit in figure, the impedance seen looking into the circuit is,

$$R_{\text{in}} = \frac{v_x}{i_x} = r_\pi + (\beta + 1)(r_O \parallel R_L)$$

(The analysis uses the relation $g_m r_\pi = (I_C / V_T)(V_T / I_B) = \beta$, which follows from the evaluation of these parameters in terms of the bias currents.) Assuming the usual case where $r_O >> R_L$, the impedance looking into the buffer is larger than the load R_L without the buffer by a factor of $(\beta + 1)$, which is substantial because β is large. The impedance is increased even more by the added r_π, but often $r_\pi << (\beta + 1) R_L$, so the addition does not make much difference

Impedance Transformation using the MOSFET Voltage Follower

Using the small-signal circuit in Figure, the impedance seen looking into the circuit is no longer R_L but instead is infinite (at low frequencies) because the MOSFET draws no current.

As frequency is increased, the parasitic capacitances of the transistors come into play and the transformed input impedance drops with frequency.

Chart of Single-transistor Amplifiers

Some configurations of single-transistor amplifier can be used as a buffer to isolate the driver from the load. For most digital applications, an NMOS voltage follower (common drain) is the preferred configuration. These amplifiers have high input impedance, which means that the digital system will not need to supply a large current.

Amplifier type	MOSFET (NMOS)	BJT (npn)	Notes
Common gate/base	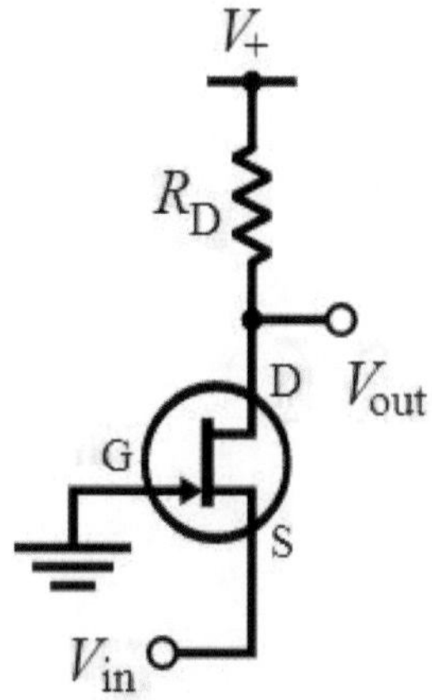	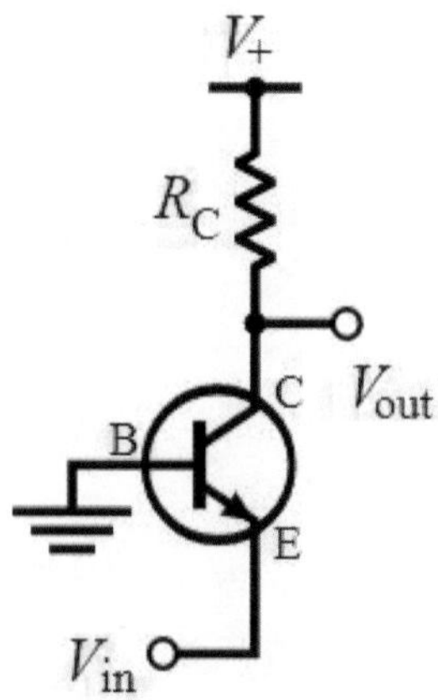	Typically used for current buffering
Common drain/ collector	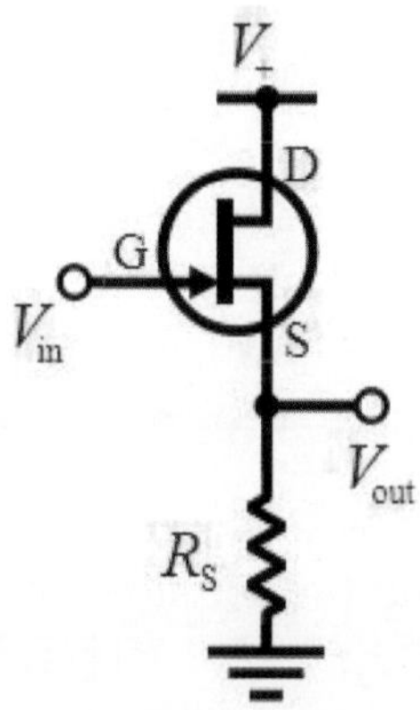	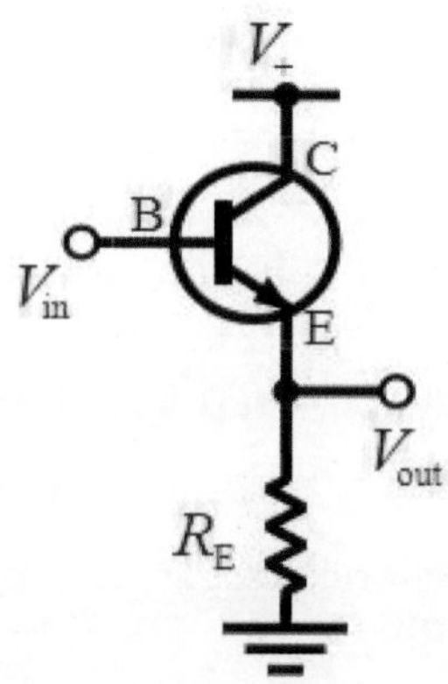	Voltage gain is close to unity, used for voltage buffering, highly dependent on the beta of a transistor.

Logic Buffer Amplifiers

A non-linear buffer amplifier is sometimes used in digital circuits where a high current is required, perhaps for driving more gates than the normal fan-out of the logic family used, or for driving displays, or long wires, or other difficult loads. It is common for a single package to contain several discrete buffer amplifiers. For example, a hex buffer is a single package containing 6 discrete buffer amplifiers, and an octal buffer is a single package containing 8 discrete buffer amplifiers. The terms inverting buffer and non-inverting buffer effectively correspond with high-current capability single-input NOR or OR gates respectively.

Speaker Array Amplifiers

The majority of amplifiers used to drive large speaker arrays, such as those used for rock concerts, are amplifiers with 26-36dB voltage gain capable of high amounts of current into low impedance speaker arrays where the speakers are wired in parallel.

Current Buffer Examples

Simple unity gain buffer amplifiers include the bipolar junction transistor in common-base configuration, or the MOSFET in common-gate configuration (called a *current follower* because the output current follows the input current). The current gain of a current buffer amplifier is (approximately) unity.

Single-transistor Circuits

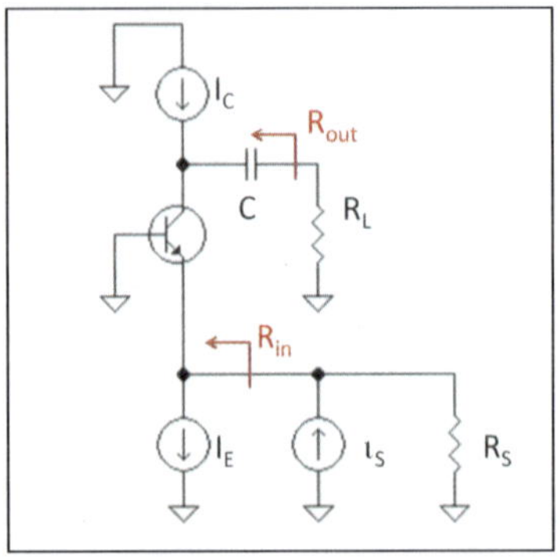

Bipolar current follower biased by current source I_E and with active load I_C.

Figure shows a bipolar current buffer biased with a current source (designated I_E for DC emitter current) and driving another DC current source as active load (designated I_C for DC collector current). The AC input signal current i_{in} is applied to the emitter node of the transistor by an AC Norton current source with Norton resistance R_S. The AC output current i_{out} is delivered by the buffer via a large coupling capacitor to load R_L. This coupling capacitor is large enough to be a short-circuit at frequencies of interest.

Because the transistor output resistance connects input and output sides of the circuit, there is a (very small) backward voltage feedback from the output to the input so this circuit is not unilateral. In addition, for the same reason, the input resistance depends (slightly) upon the output load resistance, and the output resistance depends significantly on the input driver resistance.

Current Amplifier

A Current amplifier is an electronic circuit that increases the magnitude of current of an input signal by a fixed multiple, and feeds it to the succeeding circuit/device. This process is termed as current amplification of an input signal.

The input can either be a constant signal or a time varying waveform. Ideally, during this process of current amplification, the current amplifier will keep the voltage component of the input signal unchanged. Below is the block diagram of a typical current amplifier.

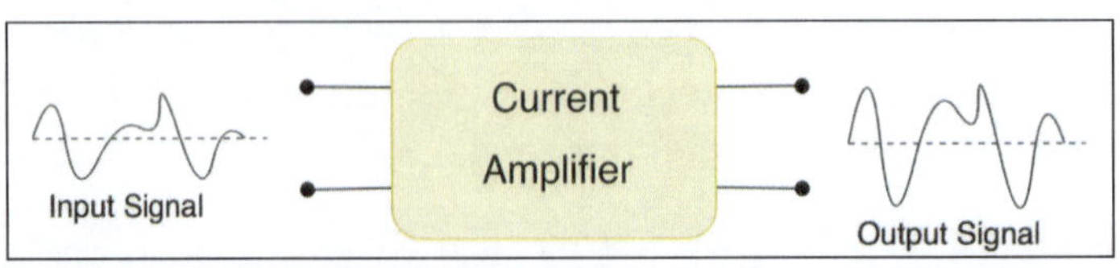

The waveforms at the input and output terminals denote the magnitude of current with respect to time. Notice that the entire waveform is stretched (increased) at the output by a fixed factor.

Gain of a Current Amplifier

In electronics, "Gain" is the technical term used for rating the amplifying capability of an amplifier. And because a current amplifier transforms only the current component of the input signal, its gain is based on how much it increases the current of output signal with respect to the input signal.

Mathematically, the gain of a current amplifier is the ratio between the magnitude of current flowing through its output terminals, to the magnitude of current of the input signal. It is denoted by the symbol Ai and because it is a ratio, it has no units.

$$\text{Current Gain } (A_i) = \frac{\text{Output Current}}{\text{Input Current}}$$

For example, if the flow of current from the input signal is 1mA and current flowing through the output terminals is 100mA, then the gain of the given current amplifier would be 100 (100mA/1mA). It means that there is a 100-fold increase in the magnitude of current of input signal at the output.

The gain can also be of negative value. It indicates that the output signal is a reversed and scaled replica of the input signal.

Characteristics of an Ideal Current Amplifier

For designing a current amplifier, a set of rules/characteristics that specify its theoretical behavior are to be made. Below are those ideal characteristics:

- The amplifiers current gain (Ai) should remain constant for the entire range of input signal;
- The current amplifiers gain should be independent of ambient conditions such as temperature and humidity;
- The input impedance (Effective resistance between the input terminals) of the current amplifier should be equal to zero;
- The output impedance (Effective resistance between the output terminals) of the current amplifier should be infinite.

In real life scenarios, the recommended impedance of current amplifiers are impossible to achieve. But they are used just as reference parameters to design near ideal current amplifier circuits. The diagram illustrates the model of an ideal current amplifier along with a practical one.

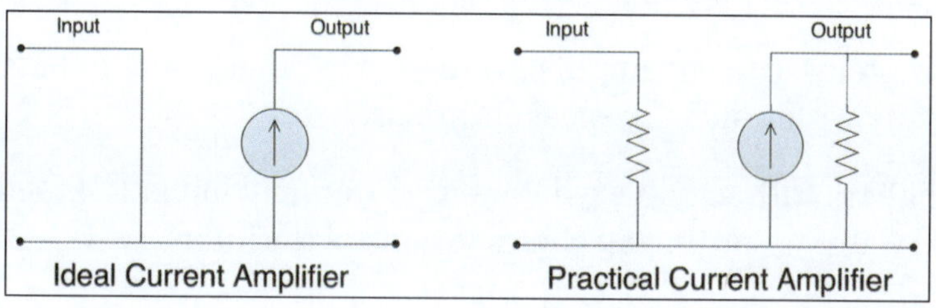

The resistances at the input and output of a current amplifier in real life scenario. The resistance in series to the input indicates the effective impedance created by the amplifying circuitry. The resistance parallel to the output denotes some of the output signal lost either by feedback mechanisms or because of internal losses.

Current Amplifier Circuit

The circuit diagram of a simple 2-stage current amplifier circuit that uses npn and pnp transistors as the amplifying element.

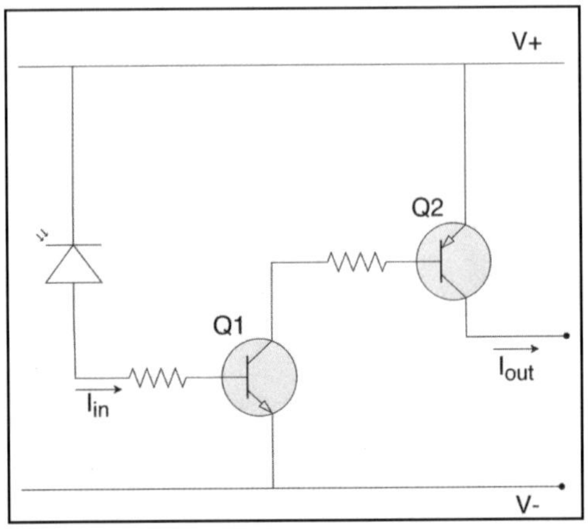

The photodiode absorbs energy from light and releases electrons, thereby acting as an input current source. This current from the photodiode is first amplified by the transistor Q1 and is further amplified by the transistor Q2.

The resistors at the bases of both the transistors are used to adjust the gain. The number of times a signal is amplified is same as stages in an amplifier. Here the current is amplified twice, so this is a 2-stage current amplifier.

Coming to the calculation part, let's say i_d is the current flowing from the photodiode and A_{i1} , A_{i2} are the gains of transistors Q1 and Q2 respectively.

The current at the output of first transistor will be equal to Ai1id, and this will be the input to second transistor. The second transistor Q2 will further amplify this signal by a factor of A_{i2}.

So, the final output current will be equal to $A_{i2}A_{i1}i_d$, effectively making gain of this entire two stage current amplifier equal to $A_{i2}A_{i1}$.

Applications of Current Amplifiers

Following are some of the practical applications of current amplifiers:

- In amplifier systems, current amplifiers are used to obtain a better bass output, by increasing the intensity with which the speakers are driven.
- Current amplifiers with variable gain are used in many industrial manufacturing systems like laser and water jet cutting machines to control the intensity with which the fabrication is done.

- In sensor systems, current amplifiers are used to strengthen weak input signals, for use in subsequent circuits.

Current Buffer

Current buffer is an electronic circuit that is used to transfer electric current from input source having very less impedance (effective resistance) to output loads with high impedance. It is designed to prevent signal sources from getting affected because of any differences in the amount of current drawn by output loads.

In most scenarios it acts as a bridge between weak input signals (like signals from sensors) and output loads that might draw larger currents. Below is the diagram of an ideal current buffer.

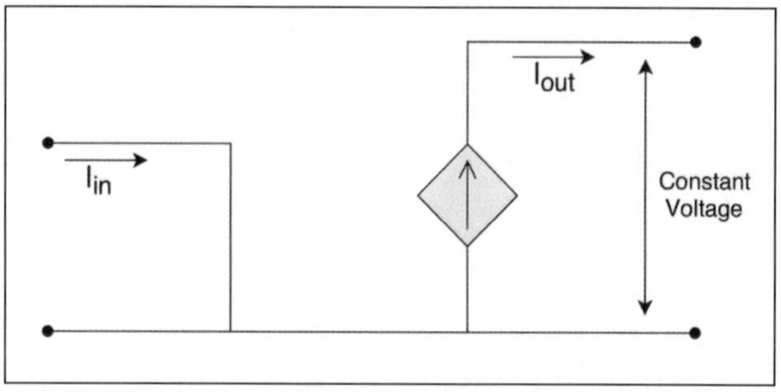

It is primarily designed to remove the influence of output load on the input source. So you can think of current buffer as a circuit that isolates input and output circuitries while allowing the required flow of current to the output load in order to maintain a constant voltage across it.

Practical use of a Current Buffer

Consider a circuit that uses an LDR sensor to drive a robot. The current consumed by the motors of robot is not constant and depends on the surface inclination or roughness i.e. load on the motors.

So, if the motors are directly coupled with the temperature sensor using a current amplifier or other similar drivers, the motors might sometimes draw more current which affects the accuracy of the sensor. The voltage across the motors will change as well, which in turn changes speed of the robot.

In order to prevent that from happening, current buffers are used. They can provide desired current to the motors without affecting accuracy of the sensor, while maintaining a constant voltage across the terminals of motors i.e. output loads.

Current Follower

A current buffer circuit with a Gain of 1 (i.e. the input and output currents are the same) is named as a current follower. It means that a current follower circuit does not provide any amplification of current to the input signal.

You might be wondering why a current follower circuit is used as the input and output currents from the current follower are the same. The reason is that a current follower not used to increase the output current.

But it is used to isolate input and output terminals while allowing the same amount of current flow into the input, and from the output. This is the reason why current follower circuits are also called as isolation buffers.

Current Buffer Circuit

Below is the circuit diagram of a simple mosfet current buffer.

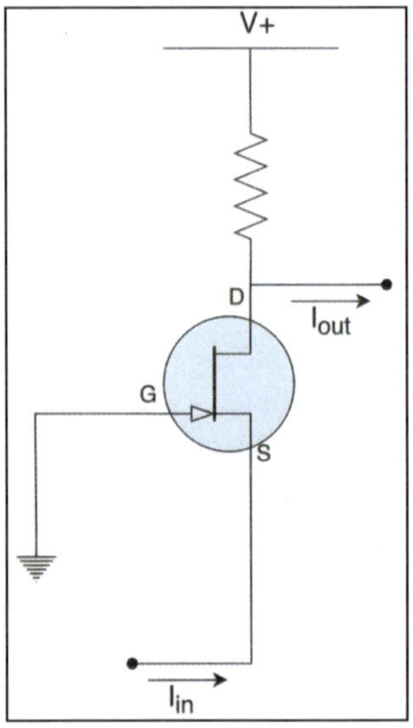

This arrangement provides very less impedance to the input signal and high impedance at the output terminal, making it a near ideal current buffer.

Applications of Current Buffers

Following are some of the practical applications of current buffers:

- In digital logic gates, current buffers are used to isolate input signals from the succeeding circuits.
- Current buffers are used in high precise sensor systems in order to reduce the influence of voltage/current fluctuations because of varying output impedances.
- In motor drivers and other electrical actuator systems.

Current Amplifiers vs. Current Buffers

At first glance, current amplifiers and current buffers may seem similar. Both take in a weak signal and produce a stronger signal (in terms of current) at the output. But the main differences lie in the manner input signal is amplified.

For current amplifiers, the magnitude of current at the output terminals is always a fixed multiple of the magnitude of input current. So given the magnitude of current of the input signal, we can easily calculate the current at the output by multiplying the input current with gain of the amplifier.

This is not the case for current buffers. The difference is that current buffers are designed to provide as much current as the load demands, while maintaining a constant voltage across it. So unless we know how much current the output load demands, we can't determine by what factor a current amplifier amplifies the input signal.

Simply put, the gain in current of a current buffer is not constant and varies depending on current demanded by the output load, while the gain of a current amplifier is constant irrespective of the output device.

Power Amplifier

A power amplifier is an electronic amplifier designed to increase the magnitude of power of a given input signal. The power of the input signal is increased to a level high enough to drive loads of output devices like speakers, headphones, RF transmitters etc. Unlike voltage/current amplifiers, a power amplifier is designed to drive loads directly and is used as a final block in an amplifier chain.

The input signal to a power amplifier needs to be above a certain threshold. So instead of directly passing the raw audio/RF signal to the power amplifier, it is first pre-amplified using current/voltage amplifiers and is sent as input to the power amp after making necessary modifications. You can observe the block diagram of an audio amplifier and the usage of power amplifier below.

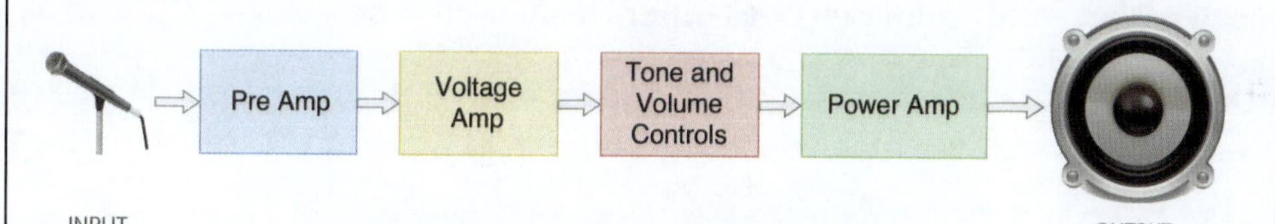

Block Diagram of an Audio Amplifier

In this case a microphone is used as an input source. The magnitude of signal from the microphone is not enough for the power amplifier. So first it is pre-amplified where its voltage and current are increased slightly. Then the signal is passed through tone and volume controls circuit which makes aesthetic adjustments to the audio waveform. Finally the signal is passed through a power amplifier and the output from power amp is fed to a speaker.

Types of Power Amplifiers

Depending on the type of output device that is connected, power amplifiers are divided into the following three types.

Audio Power Amplifiers

This type of power amplifiers are used for increasing the magnitude of power of a weaker audio

Signal. The amplifiers used in speaker driving circuitries of televisions, mobile phones etc. come under this category.

The output of an audio power amplifier ranges from a few milliwatts (like in headphone amplifiers) to thousands of watts (like power amplifiers in Hi-Fi/Home theatre systems).

Radio Frequency Power Amplifiers

Wireless transmissions require modulated waves to be sent over long distances via air. The signals are transmitted using antennas and the range of transmission depends on the magnitude of power of signals fed to the antenna.

For wireless transmissions like FM broadcasting, antennas require input signals at thousands of kilowatts of power. Here, Radio Frequency Power amplifiers are employed to increase the magnitude of power of modulated waves to a level high enough for reaching required transmission distance.

DC Power Amplifiers

DC power amplifiers are used to amplify the power of a PWM(Pulse Width Modulated) signals. They are used in electronic control systems which need high power signals to drive motors or actuators. They take input from microcontroller systems, increase its power and feed the amplified signal to DC motors or Actuators.

Power Amplifier Classes

There are multiple ways of designing a power amplifier circuit. The operation and output characteristics of each of the circuit configurations differs from each other.

To differentiate the characteristics and behaviour of different power amplifier circuits, Power Amplifier Classes are used in which letter symbols are assigned to identify the method of operation.

They are broadly classified into two categories. Power amplifiers designed to amplify analog signals come under A, B, AB or C category. Power amplifiers designed to amplify Pulse Width Modulated(PWM) digital signals come under D, E, F etc.

The most commonly used power amplifiers are the ones that are used in audio amplifier circuits and they come under classes A, B, AB or C.

Class A Power Amplifier

Analog waveforms are made up of positive highs and negative lows. In this class of amplifiers, the entire input waveform is used in the amplification process.

A single transistor is used to amplify both the positive and negative halves of the waveform. This makes their design simple and makes class A amplifiers the most commonly used type of power amplifiers. Although this class of power amplifiers are superseded by better designs, they are still popular among hobbyists.

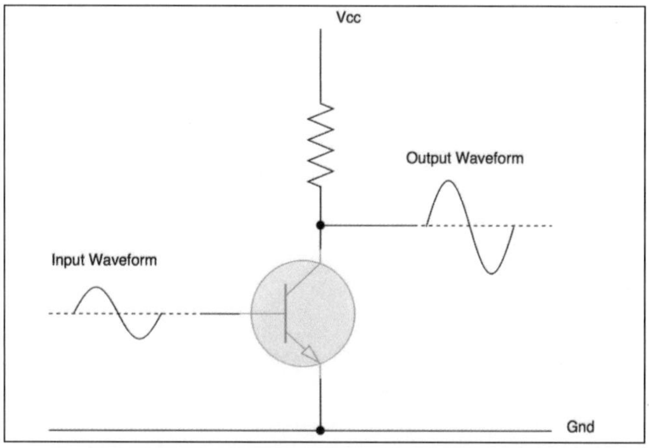

In this class of amplifiers, the active element (the electronic component used for amplifying, which is transistor in this case) is in use all the time even if there is no input signal. This generates lot of heat and reduces the efficiency of class A amplifiers to 25% in normal configuration and 50% in a transformer coupled configuration.

The conduction angle (the portion of waveform used for amplification, out of 360°) for class A amplifiers is 360°. So the signal distortion levels are very less allowing better high frequency performance.

Class B Power Amplifier

Class B power amplifiers are designed to reduce the efficiency and heating problems present in the class A amplifiers. Instead of a single transistor to amplify the entire waveform, this class of amplifiers use two complementary transistors.

One transistor amplifies positive half of the waveform and the other amplifies negative half of the waveform. So each active device conducts for one half (180°) of the waveform and two of them when combined amplify the entire signal.

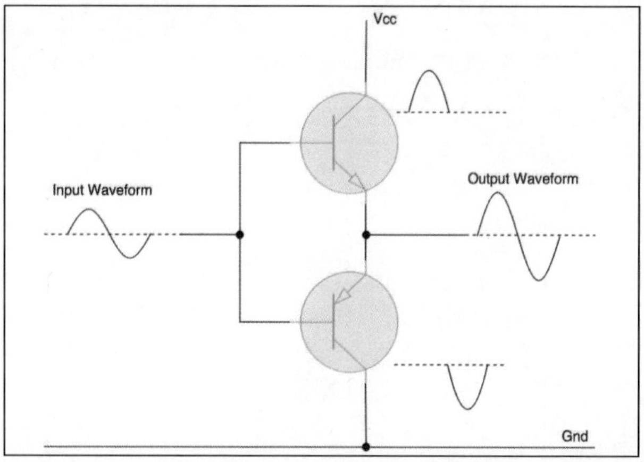

The efficiency of class B amplifiers is improved a lot over class A amplifiers because of two transistor design. They can reach a theoretical efficiency of about 75%. Power amplifiers of this class are used in battery operated devices like FM radios and transistor radios.

Because of superposition of two halves of the waveform, there exists a small distortion at the cross-over region. To reduce this signal distortion, class AB amplifiers are designed.

Class AB Power Amplifier

Class AB amplifiers are a combination of class A and class B amplifiers. This class of amplifiers are designed to reduce the less efficiency problem of class A amplifiers and distortion of signal at crossover region in class B amplifiers.

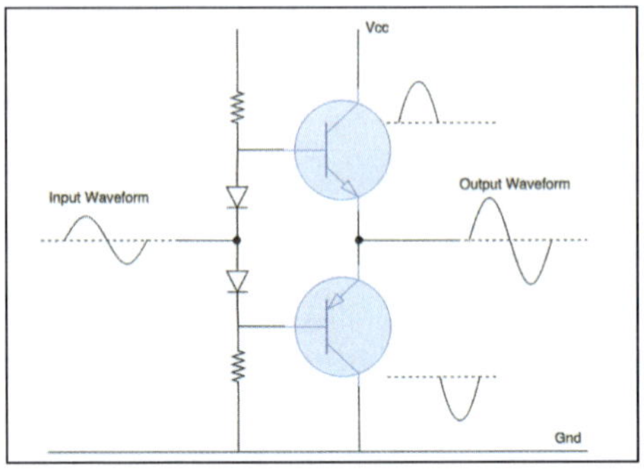

It maintains high frequency response like in class A amplifiers and good efficiency as in class B amplifiers. A combination of diodes and resistors are used to provide little bias voltage which reduces the distortion of waveform near the crossover region. There is a little drop in efficiency (60%) because of this.

Class C Power Amplifier

The design of class C power amplifiers allows greater efficiencies but reduces the linearity/conduction angle, which is under 90°. In other words, it sacrifices quality of amplification for increase in efficiency.

Lesser conduction angle implies greater distortion and so this class of amplifiers are not suited for audio amplification. They are used in high frequency oscillators and amplification of Radio Frequency signals.

Class C amplifiers generally contain a tuned load which filters and amplifies input signals of certain frequency, and the waveforms of other frequencies are supressed.

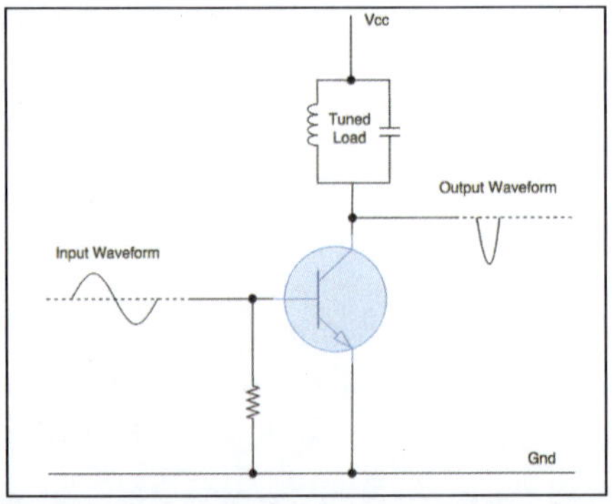

In this type of power amplifier, the active element conducts only when the input voltage is above a certain threshold, which reduces power dissipation and increases efficiency.

Other Power Amplifier Classes

Power amplifier classes D, E, F, G etc. are used to amplify PWM modulated digital signals. They come under the category of switching power amplifiers and turn the output either constantly ON or constantly OFF without any other levels in between.

Because of this simplicity, power amplifiers falling under the above mentioned classes can reach theoretical efficiencies of upto (90-100)%.

Applications

Below are the applications of power amplifiers across different sectors:

- Consumer Electronics: Audio power amplifiers are used in almost all consumer electronic devices ranging from microwave ovens, headphone drivers, televisions, mobile phones and Home theatre systems to theatrical and concert reinforcement systems.
- Industrial: Switching type power amplifiers are used for controlling most of the industrial actuator systems like servos and DC motors.
- Wireless Communication: High power amplifiers are important in transmission of cellular or FM broadcasting signals to users. Higher power levels made possible because of power amplifiers increases data transfer rates and usability. They are also used in satellite communication equipment.

Audio Power Amplifier

An audio power amplifier (or power amp) is an electronic amplifier that amplifies low-power electronic audio signals such as the signal from radio receiver or electric guitar pickup to a level that is high enough for driving loudspeakers or headphones. Audio power amplifiers are found in all manner of sound systems including sound reinforcement, public address and home audio systems and musical instrument amplifiers like guitar amplifiers. It is the final electronic stage in a typical audio playback chain before the signal is sent to the loudspeakers.

The preceding stages in such a chain are low power audio amplifiers which perform tasks like pre-amplification of the signal (this is particularly associated with record turntable signals, microphone signals and electric instrument signals from pickups, such as the electric guitar and electric bass), equalization (e.g., adjusting the bass and treble), tone controls, mixing different input signals or adding electronic effects such as reverb. The inputs can also be any number of audio sources like record players, CD players, digital audio players and cassette players. Most audio power amplifiers require these low-level inputs, which are line level.

While the input signal to an audio power amplifier, such as the signal from an electric guitar, may measure only a few hundred microwatts, its output may be a few watts for small consumer electronics devices, such as clock radios, tens or hundreds of watts for a home stereo system, several thousand watts for a nightclub's sound system or tens of thousands of watts for a large rock concert sound reinforcement system. While power amplifiers are available in standalone units, typically aimed at the hi-fi audiophile market (a niche market) of audio enthusiasts and sound reinforcement system professionals, most consumer electronics sound products, such as clock radios, boom boxes and televisions have relatively small power amplifiers that are integrated inside the chassis of the main product.

Design Parameters

Three rack-mounted audio power amplifiers used in a sound reinforcement system.

Key design parameters for audio power amplifiers are frequency response, gain, noise, and distortion. These are interdependent; increasing gain often leads to undesirable increases in noise and distortion. While negative feedback actually reduces the gain, it also reduces distortion. Most audio amplifiers are linear amplifiers operating in class AB.

Until the 1970s, most amplifiers were tube amplifiers which used vacuum tubes. During the 1970s, tube amps were increasingly replaced with transistor-based amplifiers, which were lighter in weight, more reliable, and lower maintenance. Nevertheless, there are still niche markets of consumers who continue to use tube amplifiers and tube preamplifiers in the 2010s, such as with home hi-fi enthusiasts, audio engineers and music producers (who use tube preamplifiers in studio recordings to "warm up" microphone signals) and electric guitarists, electric bassists and Hammond organ players, of whom a minority continue to use tube preamps, tube power amps and tube effects units. While hi-fi enthusiasts and audio engineers doing live sound or monitoring tracks in the studio typically seek out amplifiers with the lowest distortion, electric instrument players in genres such as blues, rock music and heavy metal music, among others, use tube amplifiers because they like the natural overdrive that tube amps produce when pushed hard.

In the 2000s, the Class-D amplifier, which is much more efficient than Class AB amplifiers, is widely used in consumer electronics audio products, bass amplifiers and sound reinforcement system gear, as Class D amplifiers are much lighter in weight and produce much less heat.

Filters and Preamplifiers

Since modern digital devices, including CD and DVD players, radio receivers and tape decks already provide a "flat" signal at line level, the preamp is not needed other than as a volume control and source selector. One alternative to a separate preamp is to simply use passive volume and switching controls, sometimes integrated into a power amplifier to form an *integrated amplifier*.

Power Output Stages

The final stage of amplification, after preamplifiers, is the output stage, where the highest demands are placed on the transistors or tubes. For this reason, the design choices made around the output device (for single-ended output stages, such as in single-ended triode amplifiers) or devices (for push-pull output stages), such as the Class of operation of the output devices is often taken as the description of the whole power amplifier. For example, a Class B amplifier will probably have just the high power output devices operating cut off for half of each cycle, while the other devices (such as differential amplifier, voltage amplifier and possibly even driver transistors) operate in Class A. In a transformerless output stage, the devices are essentially in series with the power supply and output load (such as a loudspeaker), possibly via some large capacitor and/or small resistances.

Further Developments

For some years following the introduction of solid state amplifiers, their perceived sound did not have the excellent audio quality of the best valve amplifiers. This led audiophiles to believe that "tube sound" or valve sound had an intrinsic quality due to the vacuum tube technology itself. In 1970, Matti Otala published a paper on the origin of a previously unobserved form of distortion: Transient Intermodulation Distortion (TIM), later also called Slew-Induced Distortion (SID) by others. TIM distortion was found to occur during very rapid increases in amplifier output voltage.

TIM did not appear at steady state sine tone measurements, helping to hide it from design engineers prior to 1970. Problems with TIM distortion stem from reduced open loop frequency response of solid state amplifiers. Further works of Otala and other authors found the solution for TIM distortion, including increasing slew rate, decreasing preamp frequency bandwidth, and the insertion of a lag compensation circuit in the input stage of the amplifier. In high quality modern amplifiers the open loop response is at least 20 kHz, canceling TIM distortion.

The next step in advanced design was the Baxandall Theorem, created by Peter Baxandall in England. This theorem introduced the concept of comparing the ratio between the input distortion and the output distortion of an amplifier. This new idea helped audio design engineers to better evaluate the distortion processes within an amplifier.

Applications

Pyle two-channel power amplifier.

Important applications include public address systems, theatrical and concert sound reinforcement systems, and domestic systems such as a stereo or home-theatre system. Instrument amplifiers including guitar amplifiers and electric keyboard amplifiers also use audio power amplifiers. In some cases, the power amplifier for an instrument amplifier is integrated into a single amplifier "head" which contains a preamplifier, tone controls, and electronic effects. These components may be mounted in a wooden speaker cabinet to create a "combo amplifier". Musicians with unique performance needs and/or a need for very powerful amplification may create a custom setup with separate rackmount preamplifiers, equalizers, and a power amplifier mounted in a 19" road case.

Power amplifiers are available in standalone units, which are used by hi-fi audio enthusiasts and designers of public address systems (PA systems) and sound reinforcement systems. A hi-fi user of power amplifiers may have a stereo power amplifier to drive left and right speakers and a "monoblock" single channel power amplifier to drive a subwoofer. The number of power amplifiers used in a sound reinforcement setting depends on the size of the venue. A small coffeehouse may have a single power amp driving two PA speakers. A nightclub may have several power amps for the main speakers, one or more power amps for the monitor speakers (pointing towards the band) and an additional power amp for the subwoofer. A stadium concert may have a large number of power amps mounted in racks. Most consumer electronics sound products, such as TVs, boom boxes, home cinema sound systems, Casio and Yamaha electronic keyboards, "combo" guitar amps and car stereos have power amplifiers integrated inside the chassis of the main product.

Transistor

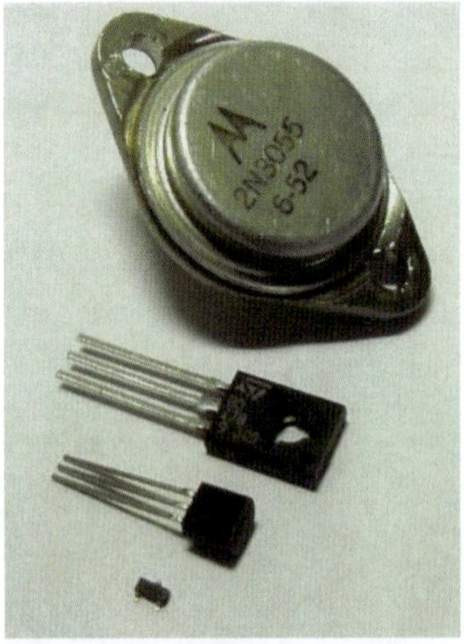

Assorted discrete transistors. Packages in order from top to bottom:
TO-3, TO-126, TO-92, SOT-23.

A transistor is a semiconductor device used to amplify or switch electronic signals and electrical power. It is composed of semiconductor material usually with at least three terminals for connection to an external circuit. A voltage or current applied to one pair of the transistor's terminals controls the current through another pair of terminals. Because the controlled (output) power can be higher than the controlling (input) power, a transistor can amplify a signal. Today, some transistors are packaged individually, but many more are found embedded in integrated circuits.

Julius Edgar Lilienfeld patented a field-effect transistor in 1926, but it was not possible to actually construct a working device at that time. The first practically implemented device was a point-contact transistor invented in 1947 by American physicists John Bardeen, Walter Brattain, and William Shockley. Bardeen, Brattain, and Shockley shared the 1956 Nobel Prize in Physics for their achievement. The most widely used transistor is the MOSFET (metal–oxide–silicon field-effect transistor), also known as the MOS transistor, which was invented by Mohamed Atalla and Dawon Kahng at Bell Labs in 1959.

Transistors revolutionized the field of electronics, and paved the way for smaller and cheaper radios, calculators, and computers, among other things. The first transistor and the MOSFET are on the list of IEEE milestones in electronics. The MOS transistor is the fundamental building block of modern electronic devices, and is ubiquitous in modern electronic systems. An estimated total of 13 sextillion MOS transistors have been manufactured as of 2018, making the MOSFET the most widely manufactured device in history.

Most transistors are made from very pure silicon, and some from germanium, but certain other semiconductor materials can also be used. A transistor may have only one kind of charge carrier, in a field-effect transistor, or may have two kinds of charge carriers in bipolar junction transistor devices. Compared with the vacuum tube, transistors are generally smaller, and require less power to operate. Certain vacuum tubes have advantages over transistors at very high operating frequencies or high operating voltages. Many types of transistors are made to standardized specifications by multiple manufacturers.

Importance

A Darlington transistor opened up so the actual transistor chip (the small square) can be seen inside. A Darlington transistor is effectively two transistors on the same chip. One transistor is much larger than the other, but both are large in comparison to transistors in large-scale integration because this particular example is intended for power applications.

The transistor is the key active component in practically all modern electronics. Many consider it to be one of the greatest inventions of the 20th century. Its importance in today's society rests on its ability to be mass-produced using a highly automated process (semiconductor device fabrication) that achieves astonishingly low per-transistor costs. The invention of the first transistor at Bell Labs was named an IEEE Milestone in 2009.

Later developments in transistor technology have also been listed among the IEEE Milestones, including the inventions of the junction transistor in 1948 and the MOSFET (metal–oxide–semiconductor field-effect transistor) in 1959. The MOSFET, also known as the MOS transistor, is the most common type of transistor in computers, electronics, and communications technology (such as smartphones). The US Patent and Trademark Office calls the MOSFET a "groundbreaking invention that transformed life and culture around the world". The MOS transistor has been the fundamental building block of modern digital electronics since the late 20th century, paving the way for the digital age.

Although several companies each produce over a billion individually packaged (known as *discrete*) transistors every year, the vast majority of transistors are now produced in integrated circuits (often shortened to *IC*, *microchips* or simply *chips*), along with diodes, resistors, capacitors and other electronic components, to produce complete electronic circuits. A logic gate consists of up to about twenty transistors whereas an advanced microprocessor, as of 2009, can use as many as 3 billion transistors (MOSFETs). "About 60 million transistors were built in 2002 for [each] man, woman, and child on Earth."

The MOS transistor is the most widely manufactured device in history. As of 2013, billions of transistors are manufactured every day, nearly all of which are MOSFET devices. As of 2018, an estimated total of 13 sextillion MOS transistors have been manufactured.

The transistor's low cost, flexibility, and reliability have made it a ubiquitous device. Transistorized mechatronic circuits have replaced electromechanical devices in controlling appliances and machinery. It is often easier and cheaper to use a standard microcontroller and write a computer program to carry out a control function than to design an equivalent mechanical system to control that same function.

Simplified Operation

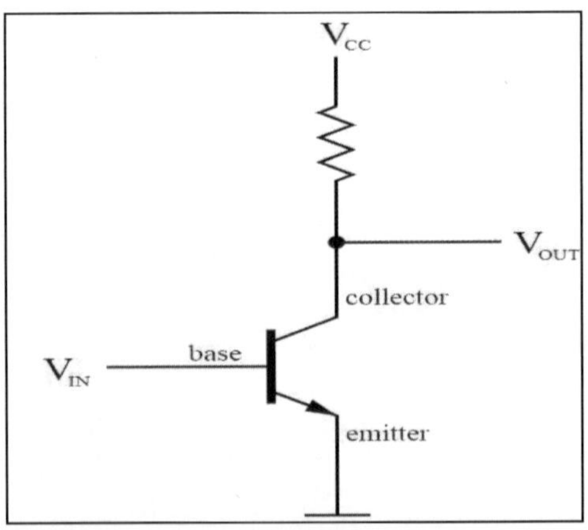

A simple circuit diagram to show the labels of a n–p–n bipolar transistor.

The image represents a typical bipolar transistor in a circuit. Charge will flow between emitter and collector terminals depending on the current in the base. Because internally the base and emitter connections behave like a semiconductor diode, a voltage drop develops between base and emitter while the base current exists. The amount of this voltage depends on the material the transistor is made from, and is referred to as V_{BE}.

The essential usefulness of a transistor comes from its ability to use a small signal applied between one pair of its terminals to control a much larger signal at another pair of terminals. This property is called gain. It can produce a stronger output signal, a voltage or current, which is proportional to a weaker input signal; that is, it can act as an amplifier. Alternatively, the transistor can be used to turn current on or off in a circuit as an electrically controlled switch, where the amount of current is determined by other circuit elements.

There are two types of transistors, which have slight differences in how they are used in a circuit. A *bipolar transistor* has terminals labeled base, collector, and emitter. A small current at the base terminal (that is, flowing between the base and the emitter) can control or switch a much larger current between the collector and emitter terminals. For a *field-effect transistor*, the terminals are labeled gate, source, and drain, and a voltage at the gate can control a current between source and drain.

Transistor as a Switch

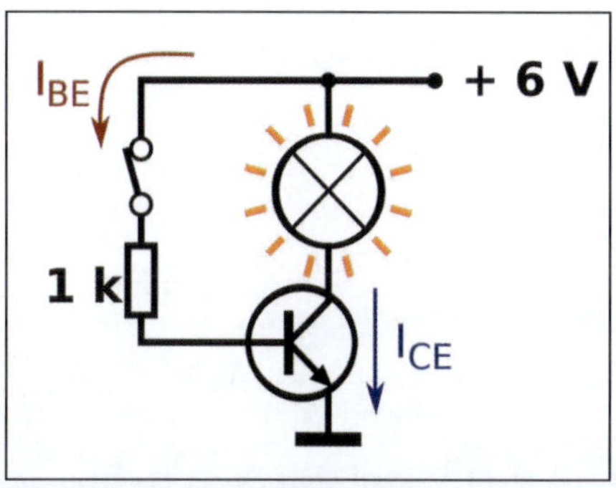

BJT used as an electronic switch, in grounded-emitter configuration.

Transistors are commonly used in digital circuits as electronic switches which can be either in an "on" or "off" state, both for high-power applications such as switched-mode power supplies and for low-power applications such as logic gates. Important parameters for this application include the current switched, the voltage handled, and the switching speed, characterised by the rise and fall times.

In a grounded-emitter transistor circuit, such as the light-switch circuit shown, as the base voltage rises, the emitter and collector currents rise exponentially. The collector voltage drops because of reduced resistance from collector to emitter. If the voltage difference between the collector and emitter were zero (or near zero), the collector current would be limited only by the load resistance (light bulb) and the supply voltage. This is called *saturation* because current is flowing from collector to emitter freely. When saturated, the switch is said to be *on*.

Providing sufficient base drive current is a key problem in the use of bipolar transistors as switches. The transistor provides current gain, allowing a relatively large current in the collector to be switched by a much smaller current into the base terminal. The ratio of these currents varies depending on the type of transistor, and even for a particular type, varies depending on the collector current. In the example light-switch circuit shown, the resistor is chosen to provide enough base current to ensure the transistor will be saturated.

In a switching circuit, the idea is to simulate, as near as possible, the ideal switch having the properties of open circuit when off, short circuit when on, and an instantaneous transition between the two states. Parameters are chosen such that the "off" output is limited to leakage currents too small to affect connected circuitry, the resistance of the transistor in the "on" state is too small to affect circuitry, and the transition between the two states is fast enough not to have a detrimental effect.

Transistor as an Amplifier

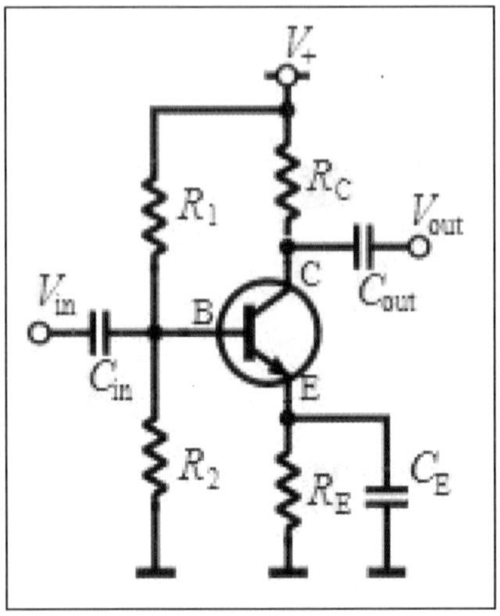

Amplifier circuit, common-emitter configuration with a voltage-divider bias circuit.

The common-emitter amplifier is designed so that a small change in voltage (V_{in}) changes the small current through the base of the transistor; the transistor's current amplification combined with the properties of the circuit means that small swings in V_{in} produce large changes in V_{out}.

Various configurations of single transistor amplifier are possible, with some providing current gain, some voltage gain, and some both.

From mobile phones to televisions, vast numbers of products include amplifiers for sound reproduction, radio transmission, and signal processing. The first discrete-transistor audio amplifiers barely supplied a few hundred milliwatts, but power and audio fidelity gradually increased as better transistors became available and amplifier architecture evolved.

Modern transistor audio amplifiers of up to a few hundred watts are common and relatively inexpensive.

Comparison with Vacuum Tubes

Before transistors were developed, vacuum (electron) tubes (or in the UK "thermionic valves" or just "valves") were the main active components in electronic equipment.

Advantages

The key advantages that have allowed transistors to replace vacuum tubes in most applications are:

- No cathode heater (which produces the characteristic orange glow of tubes), reducing power consumption, eliminating delay as tube heaters warm up, and immune from cathode poisoning and depletion;

- Very small size and weight, reducing equipment size;
- Large numbers of extremely small transistors can be manufactured as a single integrated circuit;
- Low operating voltages compatible with batteries of only a few cells;
- Circuits with greater energy efficiency are usually possible. For low-power applications (e.g., voltage amplification) in particular, energy consumption can be very much less than for tubes;
- Complementary devices available, providing design flexibility including complementary-symmetry circuits, not possible with vacuum tubes;
- Very low sensitivity to mechanical shock and vibration, providing physical ruggedness and virtually eliminating shock-induced spurious signals (e.g., microphonics in audio applications);
- Not susceptible to breakage of a glass envelope, leakage, outgassing, and other physical damage.

Limitations

Transistors have the following limitations:

- They lack the higher electron mobility afforded by the vacuum of vacuum tubes, which is desirable for high-power, high-frequency operation — such as that used in over-the-air television broadcasting;
- Transistors and other solid-state devices are susceptible to damage from very brief electrical and thermal events, including electrostatic discharge in handling; vacuum tubes are electrically much more rugged;
- They are sensitive to radiation and cosmic rays (special radiation-hardened chips are used for spacecraft devices);
- In audio applications, transistors lack the lower-harmonic distortion — the so-called tube sound — which is characteristic of vacuum tubes, and is preferred by some.

Bipolar Junction Transistor

A bipolar junction transistor (bipolar transistor or BJT) is a type of transistor that uses both electrons and holes as charge carriers.

Unipolar transistors, such as field-effect transistors, only use one kind of charge carrier. BJTs use two junctions between two semiconductor types, n-type and p-type.

BJTs are manufactured in two types: NPN and PNP, and are available as individual components, or fabricated in integrated circuits, often in large numbers.

Usage

BJTs can be used as amplifiers or switches. This ability gives them many applications in electronic equipment such as computers, televisions, mobile phones, audio amplifiers, industrial control, and radio transmitters.

Current Direction Conventions

By convention, the direction of current on diagrams is shown as the direction that a positive charge would move. This is called *conventional current*. However, current in many metal conductors is due to the flow of electrons. Because electrons carry a negative charge, they move in the direction opposite to conventional current. On the other hand, inside a bipolar transistor, currents can be composed of both positively charged holes and negatively charged electrons. current arrows are shown in the conventional direction, but labels for the movement of holes and electrons show their actual direction inside the transistor. The arrow on the symbol for bipolar transistors indicates the PN junction between base and emitter and points in the direction in which conventional current travels.

Function

BJTs are available in two types, or polarities, known as PNP and NPN based on the doping types of the three main terminal regions. An NPN transistor comprises two semiconductor junctions that share a thin p-doped region, and a PNP transistor comprises two semiconductor junctions that share a thin n-doped region.

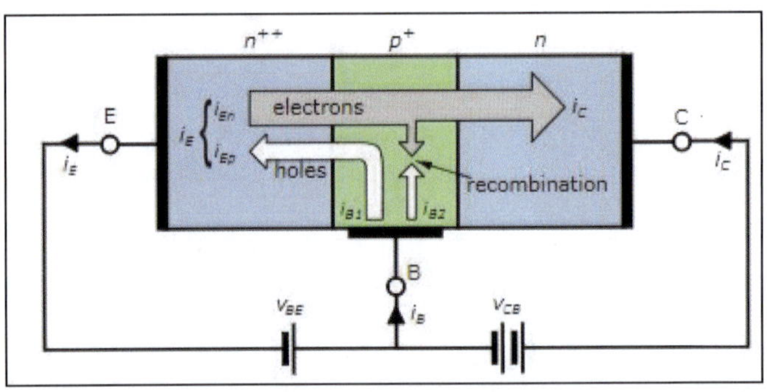

NPN BJT with forward-biased E–B junction and reverse-biased B–C junction.

Charge flow in a BJT is due to diffusion of charge carriers across a junction between two regions of different charge concentrations. The regions of a BJT are called *emitter*, *base*, and *collector*. A discrete transistor has three leads for connection to these regions. Typically, the emitter region is heavily doped compared to the other two layers, and the collector is doped much lighter than the base (collector doping is typically ten times lighter than base doping). By design, most of the BJT collector current is due to the flow of charge carriers (electrons or holes) injected from a heavily doped emitter into the base where they are minority carriers that diffuse toward the collector, and so BJTs are classified as minority-carrier devices.

In typical operation, the base–emitter junction is forward-biased, which means that the p-doped side of the junction is at a more positive potential than the n-doped side, and the base–collector junction is reverse-biased. When forward bias is applied to the base–emitter junction, the equilibrium between the thermally generated carriers and the repelling electric field of the n-doped emitter depletion region is disturbed. This allows thermally excited electrons to inject from the emitter into the base region. These electrons diffuse through the base from the region of high concentration near the emitter toward the region of low concentration near the collector. The electrons in the base are called *minority carriers* because the base is doped p-type, which makes holes the *majority carrier* in the base.

To minimize the fraction of carriers that recombine before reaching the collector–base junction, the transistor's base region must be thin enough that carriers can diffuse across it in much less time than the semiconductor's minority-carrier lifetime. Having a lightly doped base ensures recombination rates are low. In particular, the thickness of the base must be much less than the diffusion length of the electrons. The collector–base junction is reverse-biased, and so negligible electron injection occurs from the collector to the base, but carriers that are injected into the base and diffuse to reach the collector-base depletion region are swept into the collector by the electric field in the depletion region. The thin *shared* base and asymmetric collector–emitter doping are what differentiates a bipolar transistor from two *separate* and oppositely biased diodes connected in series.

Voltage, Current, and Charge Control

The collector–emitter current can be viewed as being controlled by the base–emitter current (current control), or by the base–emitter voltage (voltage control). These views are related by the current–voltage relation of the base–emitter junction, which is the usual exponential current–voltage curve of a p–n junction (diode).

The explanation for collector current is the concentration gradient of minority carriers in the base region. Due to low-level injection (in which there are much fewer excess carriers than normal majority carriers) the ambipolar transport rates (in which the excess majority and minority carriers flow at the same rate) is in effect determined by the excess minority carriers.

Detailed transistor models of transistor action, such as the Gummel–Poon model, account for the distribution of this charge explicitly to explain transistor behaviour more exactly. The charge-control view easily handles phototransistors, where minority carriers in the base region are created by the absorption of photons, and handles the dynamics of turn-off, or recovery time, which depends on charge in the base region recombining. However, because base charge is not a signal that is visible at the terminals, the current- and voltage-control views are generally used in circuit design and analysis.

In analog circuit design, the current-control view is sometimes used because it is approximately linear. That is, the collector current is approximately β_F times the base current. Some basic circuits can be designed by assuming that the base-emitter voltage is approximately constant and that collector current is β times the base current. However, to accurately and reliably design production BJT circuits, the voltage-control (for example, Ebers–Moll) model is required. The voltage-control model requires an exponential function to be taken into account, but when it is linearized such that the transistor can be modeled as a transconductance, as in the Ebers–Moll model, design for circuits such as differential amplifiers again becomes a mostly linear problem, so the voltage-control view is often preferred. For translinear circuits, in which the exponential I–V curve is key to the operation, the transistors are usually modeled as voltage-controlled current sources whose transconductance is proportional to their collector current. In general, transistor-level circuit analysis is performed using SPICE or a comparable analog-circuit simulator, so mathematical model complexity is usually not of much concern to the designer, but a simplified view of the characteristics allows designs to be created following a logical process.

Turn-on, Turn-off, and Storage Delay

Bipolar transistors, and particularly power transistors, have long base-storage times when they are driven into saturation; the base storage limits turn-off time in switching applications. A Baker clamp can prevent the transistor from heavily saturating, which reduces the amount of charge stored in the base and thus improves switching time.

Transistor Characteristics: Alpha (α) and Beta (β)

The proportion of carriers able to cross the base and reach the collector is a measure of the BJT efficiency. The heavy doping of the emitter region and light doping of the base region causes many more electrons to be injected from the emitter into the base than holes to be injected from the base into the emitter. A thin and lightly-doped base region means that most of the minority carriers that are injected into the base will diffuse to the collector and not recombine.

The *common-emitter current gain* is represented by β_F or the *h*-parameter h_{FE}; it is approximately the ratio of the DC collector current to the DC base current in forward-active region. It is typically greater than 50 for small-signal transistors, but can be smaller in transistors designed for high-power applications. Both injection efficiency and recombination in the base reduce the BJT gain.

Another useful characteristic is the *common-base current gain*, α_F. The common-base current gain is approximately the gain of current from emitter to collector in the forward-active region. This ratio usually has a value close to unity; between 0.980 and 0.998. It is less than unity due to recombination of charge carriers as they cross the base region.

Alpha and beta are related by the following identities:

$$\alpha_F = \frac{I_C}{I_E}, \qquad \beta_F = \frac{I_C}{I_B},$$

$$\alpha_F = \frac{\beta_F}{1+\beta_F} \qquad \Leftrightarrow \beta_F = \frac{\alpha_F}{1-\alpha_F}.$$

Beta is a convenient figure of merit to describe the performance of a bipolar transistor, but is not a fundamental physical property of the device. Bipolar transistors can be considered voltage-controlled devices (fundamentally the collector current is controlled by the base-emitter voltage; the base current could be considered a defect and is controlled by the characteristics of the base-emitter junction and recombination in the base). In many designs beta is assumed high enough so that base current has a negligible effect on the circuit. In some circuits (generally switching circuits), sufficient base current is supplied so that even the lowest beta value a particular device may have will still allow the required collector current to flow.

Structure

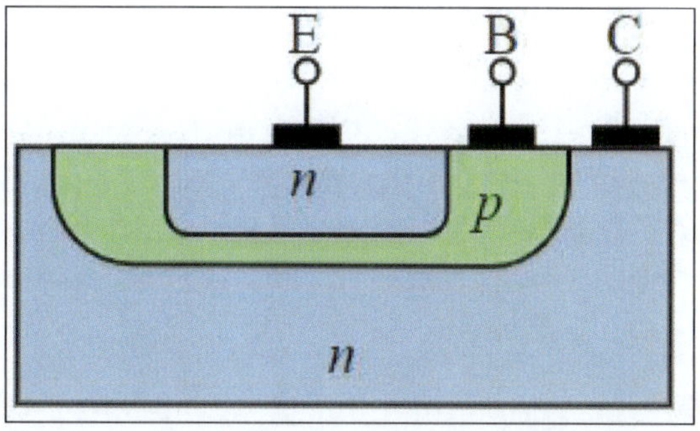

Simplified cross section of a planar *NPN* bipolar junction transistor.

A BJT consists of three differently doped semiconductor regions: the *emitter* region, the *base* region and the *collector* region. These regions are, respectively, *p* type, *n* type and *p* type in a PNP transistor, and *n* type, *p* type and *n* type in an NPN transistor. Each semiconductor region is connected to a terminal, appropriately labeled: *emitter* (E), *base* (B) and *collector* (C).

The *base* is physically located between the *emitter* and the *collector* and is made from lightly doped, high-resistivity material. The collector surrounds the emitter region, making it almost impossible for the electrons injected into the base region to escape without being collected, thus making the resulting value of α very close to unity, and so, giving the transistor a large β. A cross-section view of a BJT indicates that the collector–base junction has a much larger area than the emitter–base junction.

The bipolar junction transistor, unlike other transistors, is usually not a symmetrical device. This means that interchanging the collector and the emitter makes the transistor leave the forward active mode and start to operate in reverse mode. Because the transistor's internal structure is usually optimized for forward-mode operation, interchanging the collector and the emitter makes the values of α and β in reverse operation much smaller than those in forward operation; often the α of the reverse mode is lower than 0.5. The lack of symmetry is primarily due to the doping ratios of the emitter and the collector. The emitter is heavily doped, while the collector is lightly doped, allowing a large reverse bias voltage to be applied before the collector–base junction breaks down. The collector–base junction is reverse biased in normal operation. The reason the emitter is heavily doped is to increase the emitter injection efficiency: the ratio of carriers injected by the emitter to those injected by the base. For high current gain, most of the carriers injected into the emitter–base junction must come from the emitter.

Die of a KSY34 high-frequency NPN transistor. Bond wires connect to the base and emitter.

The low-performance "lateral" bipolar transistors sometimes used in CMOS processes are sometimes designed symmetrically, that is, with no difference between forward and backward operation.

Small changes in the voltage applied across the base–emitter terminals cause the current between the *emitter* and the *collector* to change significantly. This effect can be used to amplify the input voltage or current. BJTs can be thought of as voltage-controlled current sources, but are more simply characterized as current-controlled current sources, or current amplifiers, due to the low impedance at the base.

Early transistors were made from germanium but most modern BJTs are made from silicon. A significant minority are also now made from gallium arsenide, especially for very high speed applications.

NPN

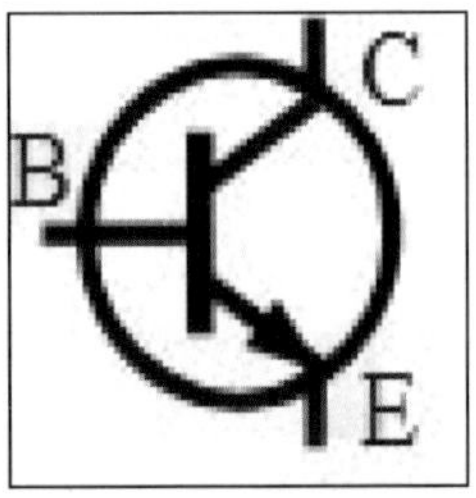

The symbol of an NPN BJT. A mnemonic for the symbol is "not pointing in".

NPN is one of the two types of bipolar transistors, consisting of a layer of P-doped semiconductor (the "base") between two N-doped layers. A small current entering the base is amplified to produce a large collector and emitter current. That is, when there is a positive potential difference measured from the base of an NPN transistor to its emitter (that is, when the base is high relative to the emitter), as well as a positive potential difference measured from the collector to the emitter, the transistor becomes active. In this "on" state, current flows from the collector to the emitter of the transistor. Most of the current is carried by electrons moving from emitter to collector as minority carriers in the P-type base region. To allow for greater current and faster operation, most bipolar transistors used today are NPN because electron mobility is higher than hole mobility.

PNP

The symbol of a PNP BJT. A mnemonic for the symbol is "points in proudly".

The other type of BJT is the PNP, consisting of a layer of N-doped semiconductor between two layers of P-doped material. A small current leaving the base is amplified in the collector output. That is, a PNP transistor is "on" when its base is pulled low relative to the emitter. In a PNP transistor, the emitter–base region is forward biased, so holes are injected into the base as minority carriers. The base is very thin, and most of the holes cross the reverse-biased base–collector junction to the collector.

The arrows in the NPN and PNP transistor symbols indicate the PN junction between the base and emitter. When the device is in forward active or forward saturated mode, the arrow, placed on the emitter leg, points in the direction of the conventional current.

Heterojunction Bipolar Transistor

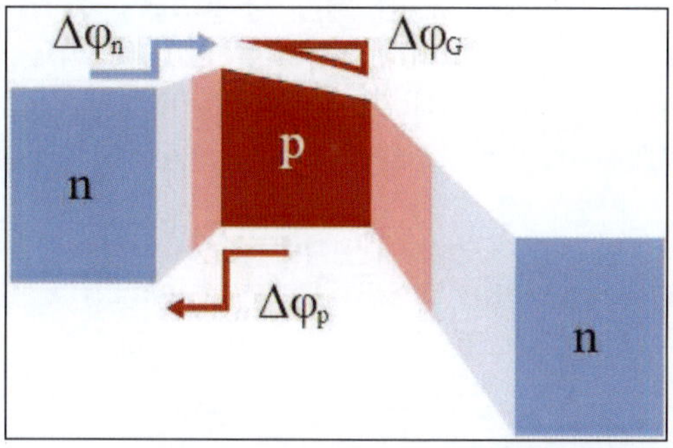

Bands in graded heterojunction NPN bipolar transistor. Barriers indicated for electrons to move from emitter to base and for holes to be injected backward from base to emitter; also, grading of bandgap in base assists electron transport in base region. Light colors indicate depleted regions.

The heterojunction bipolar transistor (HBT) is an improvement of the BJT that can handle signals of very high frequencies up to several hundred GHz. It is common in modern ultrafast circuits, mostly RF systems.

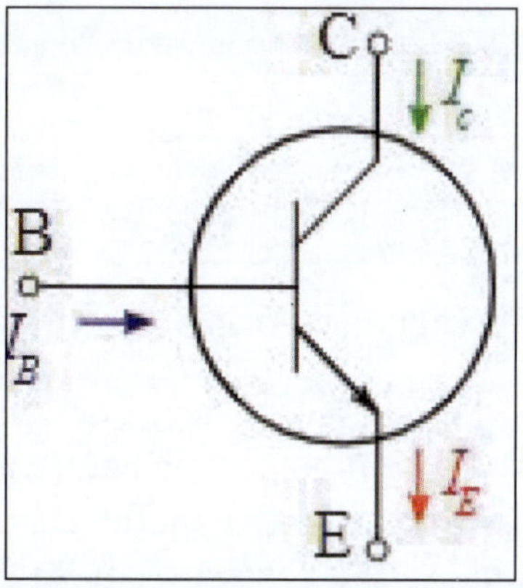

Symbol for NPN bipolar transistor with current flow direction.

Heterojunction transistors have different semiconductors for the elements of the transistor. Usually the emitter is composed of a larger bandgap material than the base. The figure shows that this difference in bandgap allows the barrier for holes to inject backward from the base into the emitter, denoted in the figure as $\Delta\varphi_p$, to be made large, while the barrier for electrons to inject into the base $\Delta\varphi_n$ is made low. This barrier arrangement helps reduce minority carrier injection from the base when the emitter-base junction is under forward bias, and thus reduces base current and increases emitter injection efficiency.

The improved injection of carriers into the base allows the base to have a higher doping level, resulting in lower resistance to access the base electrode. In the more traditional BJT, also referred to as homojunction BJT, the efficiency of carrier injection from the emitter to the base is primarily determined by the doping ratio between the emitter and base, which means the base must be lightly doped to obtain high injection efficiency, making its resistance relatively high. In addition, higher doping in the base can improve figures of merit like the early voltage by lessening base narrowing.

The grading of composition in the base, for example, by progressively increasing the amount of germanium in a SiGe transistor, causes a gradient in bandgap in the neutral base, denoted in the figure by $\Delta\varphi_G$, providing a "built-in" field that assists electron transport across the base. That drift component of transport aids the normal diffusive transport, increasing the frequency response of the transistor by shortening the transit time across the base.

Two commonly used HBTs are silicon–germanium and aluminum gallium arsenide, though a wide variety of semiconductors may be used for the HBT structure. HBT structures are usually grown by epitaxy techniques like MOCVD and MBE.

Regions of Operation

Junction type	Applied voltages	Junction bias		Mode
		B-E	B-C	
NPN	E < B < C	Forward	Reverse	Forward-active
	E < B > C	Forward	Forward	Saturation
	E > B < C	Reverse	Reverse	Cut-off
	E > B > C	Reverse	Forward	Reverse-active
PNP	E < B < C	Reverse	Forward	Reverse-active
	E < B > C	Reverse	Reverse	Cut-off
	E > B < C	Forward	Forward	Saturation
	E > B > C	Forward	Reverse	Forward-active

Bipolar transistors have four distinct regions of operation, defined by BJT junction biases.

Forward-active (or simply *active*): The base–emitter junction is forward biased and the base–collector junction is reverse biased. Most bipolar transistors are designed to afford the greatest common-emitter current gain, β_F, in forward-active mode. If this is the case, the collector–emitter current is approximately proportional to the base current, but many times larger, for small base current variations.

Reverse-active (or *inverse-active* or *inverted*): By reversing the biasing conditions of the forward-active region, a bipolar transistor goes into reverse-active mode. In this mode, the emitter and collector regions switch roles. Because most BJTs are designed to maximize current gain in forward-active mode, the β_F in inverted mode is several times smaller (2–3 times for the ordinary germanium transistor). This transistor mode is seldom used, usually being considered only for failsafe conditions and some types of bipolar logic. The reverse bias breakdown voltage to the base may be an order of magnitude lower in this region.

Saturation: With both junctions forward-biased, a BJT is in saturation mode and facilitates high current conduction from the emitter to the collector (or the other direction in the case of NPN, with negatively charged carriers flowing from emitter to collector). This mode corresponds to a logical "on", or a closed switch.

Cut-off: In cut-off, biasing conditions opposite of saturation (both junctions reverse biased) are present. There is very little current, which corresponds to a logical "off", or an open switch.

Avalanche breakdown region:

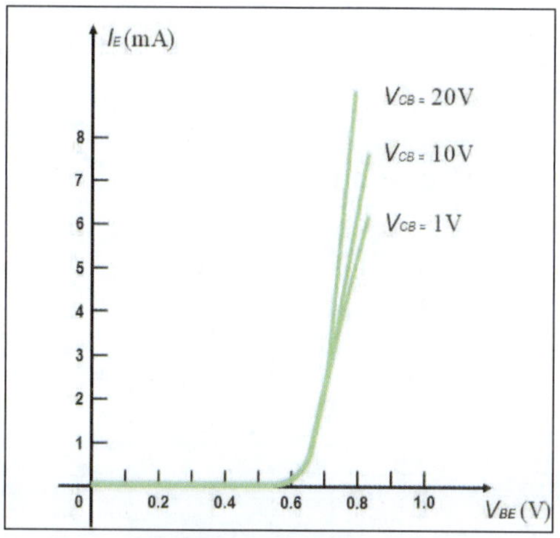

Input characteristics.

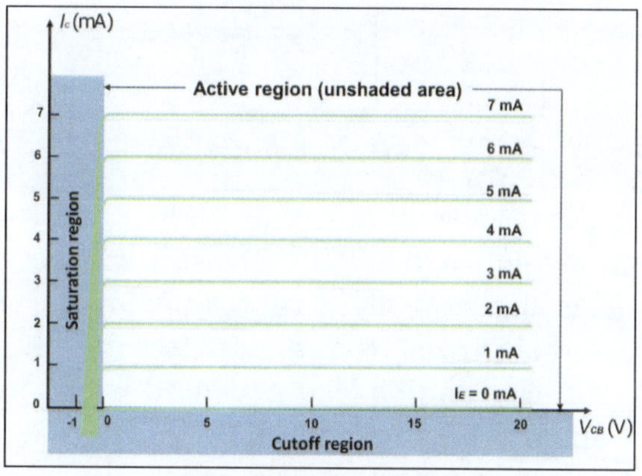

Output characteristics.

Input and output characteristics for a common-base silicon transistor amplifier.

The modes of operation can be described in terms of the applied voltages (this description applies to NPN transistors; polarities are reversed for PNP transistors).

- Forward-active: Base higher than emitter, collector higher than base (in this mode the collector current is proportional to base current by β_F).
- Saturation: Base higher than emitter, but collector is not higher than base.
- Cut-off: Base lower than emitter, but collector is higher than base. It means the transistor is not letting conventional current go through from collector to emitter.

- Reverse-active: Base lower than emitter, collector lower than base: reverse conventional current goes through transistor.
- In terms of junction biasing: (*reverse biased base–collector junction* means $V_{bc} < 0$ for NPN, opposite for PNP)

Although these regions are well defined for sufficiently large applied voltage, they overlap somewhat for small (less than a few hundred millivolts) biases. For example, in the typical grounded-emitter configuration of an NPN BJT used as a pulldown switch in digital logic, the "off" state never involves a reverse-biased junction because the base voltage never goes below ground; nevertheless the forward bias is close enough to zero that essentially no current flows, so this end of the forward active region can be regarded as the cutoff region.

Active-mode Transistors in Circuits

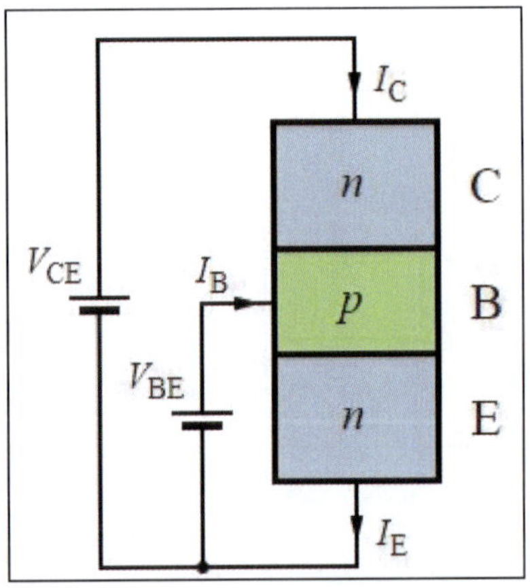

Structure and use of NPN transistor. Arrow according to schematic.

The diagram shows a schematic representation of an NPN transistor connected to two voltage sources. (The same description applies to a PNP transistor with reversed directions of current flow and applied voltage.) This applied voltage causes the lower P-N junction to become forward biased, allowing a flow of electrons from the emitter into the base. In active mode, the electric field existing between base and collector (caused by V_{CE}) will cause the majority of these electrons to cross the upper P-N junction into the collector to form the collector current I_C. The remainder of the electrons recombine with holes, the majority carriers in the base, making a current through the base connection to form the base current, I_B. As shown in the diagram, the emitter current, I_E, is the total transistor current, which is the sum of the other terminal currents, (i.e., $I_E = I_B + I_C$).

In the diagram, the arrows representing current point in the direction of conventional current – the flow of electrons is in the opposite direction of the arrows because electrons carry negative electric charge. In active mode, the ratio of the collector current to the base current is called the *DC current gain*. This gain is usually 100 or more, but robust circuit designs do not depend on the exact value. The value of this gain for DC signals is referred to as h_{FE}, and the value of this gain for small signals is referred to as h_{fe}. That is, when a small change in the currents occurs, and sufficient time has passed for the new condition to reach a steady state h_{fe} is the ratio of the change in collector current to the change in base current. The symbol β is used for both h_{FE} and h_{fe}.

The emitter current is related to V_{BE} exponentially. At room temperature, an increase in V_{BE} by approximately 60 mV increases the emitter current by a factor of 10. Because the base current is approximately proportional to the collector and emitter currents, they vary in the same way.

Theory and Modeling

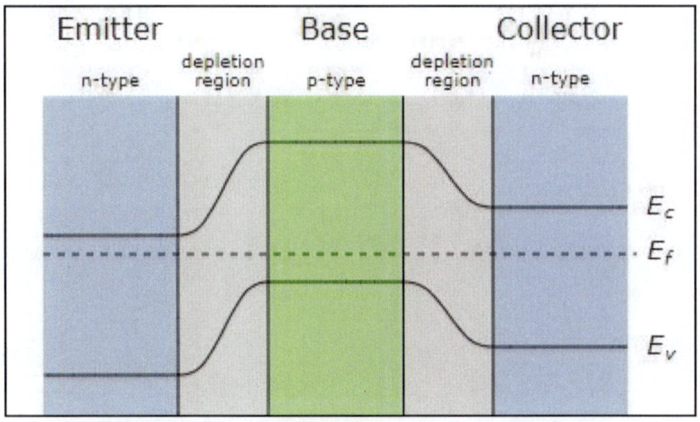

Band diagram for NPN transistor at equilibrium.

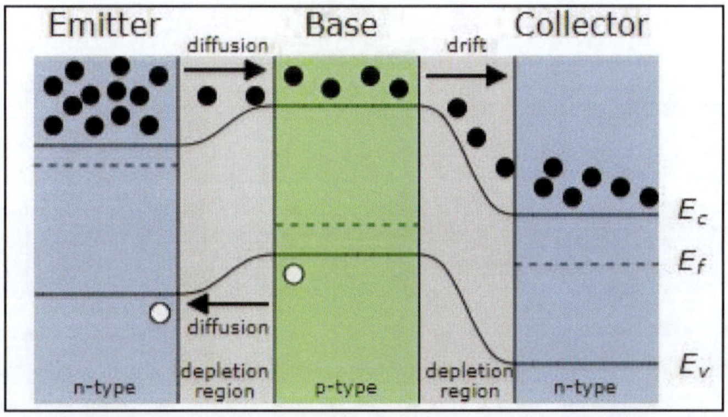

Band diagram for NPN transistor in active mode, showing injection of electrons from emitter to base, and their overshoot into the collector.

Transistors can be thought of as two diodes (P–N junctions) sharing a common region that minority carriers can move through. A PNP BJT will function like two diodes that share an N-type cathode region, and the NPN like two diodes sharing a P-type anode region. Connecting two diodes with wires will not make a transistor, since minority carriers will not be able to get from one P–N junction to the other through the wire.

Both types of BJT function by letting a small current input to the base control an amplified output from the collector. The result is that the transistor makes a good switch that is controlled by its base input. The BJT also makes a good amplifier, since it can multiply a weak input signal to about 100 times its original strength. Networks of transistors are used to make powerful amplifiers with many different applications. Focus is on the NPN bipolar transistor. In the NPN transistor in what is called active mode, the base–emitter voltage V_{BE} and collector–base voltage V_{CB} are positive, forward biasing the emitter–base junction and reverse-biasing the collector–base junction. In the active mode of operation, electrons are injected from the forward biased n-type emitter region into the p-type base where they diffuse as minority carriers to the reverse-biased n-type collector and are swept away by the electric field in the reverse-biased collector–base junction.

Large-signal Models

In 1954, Jewell James Ebers and John L. Moll introduced their mathematical model of transistor currents.

Ebers–Moll Model

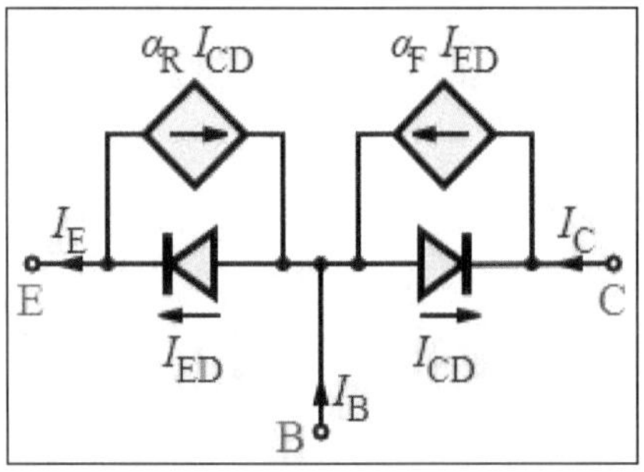

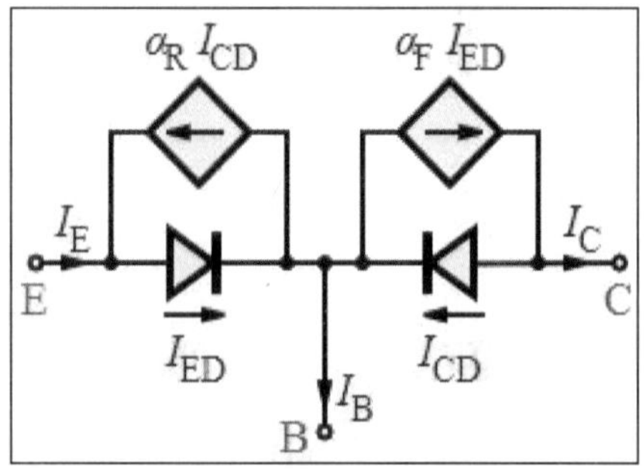

Ebers–Moll model for a PNP transistor.

Ebers–Moll model for an NPN transistor. I_B, I_C and I_E are the base, collector and emitter currents; I_{CD} and I_{ED} are the collector and emitter diode currents; α_F and α_R are the forward and reverse common-base current gains.

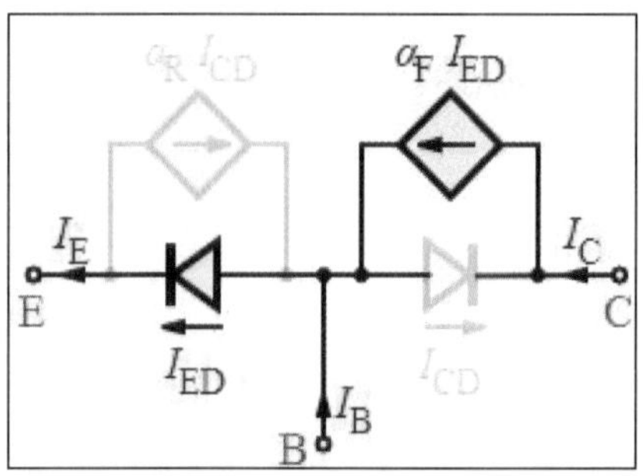

Approximated Ebers–Moll model for an NPN transistor in the forward active mode. The collector diode is reverse-biased so I_{CD} is virtually zero. Most of the emitter diode current (α_F is nearly 1) is drawn from the collector, providing the amplification of the base current.

The DC emitter and collector currents in active mode are well modeled by an approximation to the Ebers–Moll model:

$$I_E = I_{ES}\left(e^{\frac{V_{BE}}{V_T}} - 1\right)$$

$$I_C = \alpha_F I_E$$

$$I_B = (1-\alpha_F) I_E$$

The base internal current is mainly by diffusion and,

$$J_{n(\text{base})} = \frac{1}{W} q D_n n_{bo} e^{\frac{V_{EB}}{V_T}}$$

where,

- V_T is the thermal voltage kT/q (approximately 26 mV at 300 K ≈ room temperature);
- I_E is the emitter current;
- I_C is the collector current;
- α_F is the common base forward short-circuit current gain (0.98 to 0.998);
- I_{ES} is the reverse saturation current of the base–emitter diode (on the order of 10^{-15} to 10^{-12} amperes);
- V_{BE} is the base–emitter voltage;
- D_n is the diffusion constant for electrons in the p-type base;
- W is the base width.

The α and forward β parameters are as described previously. A reverse β is sometimes included in the model.

The unapproximated Ebers–Moll equations used to describe the three currents in any operating region are given below. These equations are based on the transport model for a bipolar junction transistor.

$$i_C = I_S\left[\left(e^{\frac{V_{BE}}{V_T}} - e^{\frac{V_{BC}}{V_T}}\right) - \frac{1}{\beta_R}\left(e^{\frac{V_{BC}}{V_T}} - 1\right)\right]$$

$$i_B = I_S\left[\frac{1}{\beta_F}\left(e^{\frac{V_{BE}}{V_T}} - 1\right) + \frac{1}{\beta_R}\left(e^{\frac{V_{BC}}{V_T}} - 1\right)\right]$$

$$i_E = I_S\left[\left(e^{\frac{V_{BE}}{V_T}} - e^{\frac{V_{BC}}{V_T}}\right) + \frac{1}{\beta_F}\left(e^{\frac{V_{BE}}{V_T}} - 1\right)\right]$$

where,

- i_C is the collector current;
- i_B is the base current;
- i_E is the emitter current;
- β_F is the forward common emitter current gain (20 to 500);
- is the reverse common emitter current gain (0 to 20);
- I_S is the reverse saturation current (on the order of 10^{-15} to 10^{-12} amperes);
- V_T is the thermal voltage (approximately 26 mV at 300 K ≈ room temperature);

- V_{BE} is the base–emitter voltage;
- V_{BC} is the base–collector voltage.

Base-width Modulation

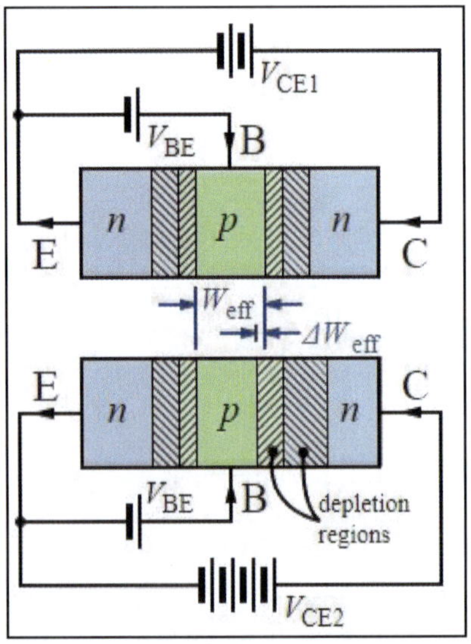

Top: NPN base width for low collector–base reverse bias; Bottom: narrower NPN base width for large collector–base reverse bias. Hashed regions are depleted regions.

As the collector–base voltage ($V_{CB} = V_{CE} - V_{BE}$) varies, the collector–base depletion region varies in size. An increase in the collector–base voltage, for example, causes a greater reverse bias across the collector–base junction, increasing the collector–base depletion region width, and decreasing the width of the base. This variation in base width often is called the *Early effect* after its discoverer James M. Early.

Narrowing of the base width has two consequences:

- There is a lesser chance for recombination within the "smaller" base region.
- The charge gradient is increased across the base, and consequently, the current of minority carriers injected across the emitter junction increases.

Both factors increase the collector or "output" current of the transistor in response to an increase in the collector–base voltage.

In the forward-active region, the Early effect modifies the collector current (i_C) and the forward common emitter current gain (β_F) as given by:

$$i_C = I_S e^{\frac{v_{BE}}{V_T}} \left(1 + \frac{V_{CE}}{V_A}\right)$$

$$\beta_F = \beta_{F0} \left(1 + \frac{V_{CB}}{V_A}\right)$$

$$r_o = \frac{V_A}{I_C}$$

where,

- V_{CE} is the collector–emitter voltage;
- V_A is the Early voltage (15 V to 150 V);
- β_{F0} is forward common-emitter current gain when $V_{CB} = 0$ V;
- r_o is the output impedance;
- I_C is the collector current.

Punchthrough

When the base–collector voltage reaches a certain (device-specific) value, the base–collector depletion region boundary meets the base–emitter depletion region boundary. When in this state the transistor effectively has no base. The device thus loses all gain when in this state.

Gummel–Poon Charge-control Model

The Gummel–Poon model is a detailed charge-controlled model of BJT dynamics, which has been adopted and elaborated by others to explain transistor dynamics in greater detail than the terminal-based models typically do. This model also includes the dependence of transistor β-values upon the direct current levels in the transistor, which are assumed current-independent in the Ebers–Moll model.

Small-signal Models

Hybrid-pi Model

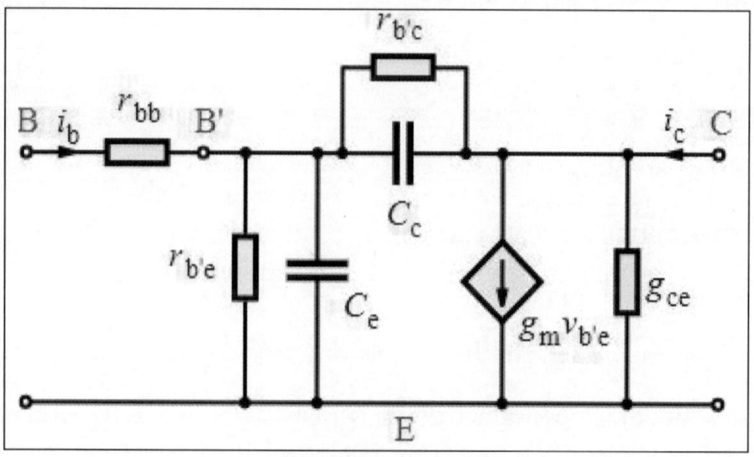

Hybrid-pi model.

The hybrid-pi model is a popular circuit model used for analyzing the small signal and AC behavior of bipolar junction and field effect transistors. Sometimes it is also called *Giacoletto model* because it was introduced by L.J. Giacoletto in 1969. The model can be quite accurate for low-frequency circuits and can easily be adapted for higher-frequency circuits with the addition of appropriate inter-electrode capacitances and other parasitic elements.

h-parameter Model

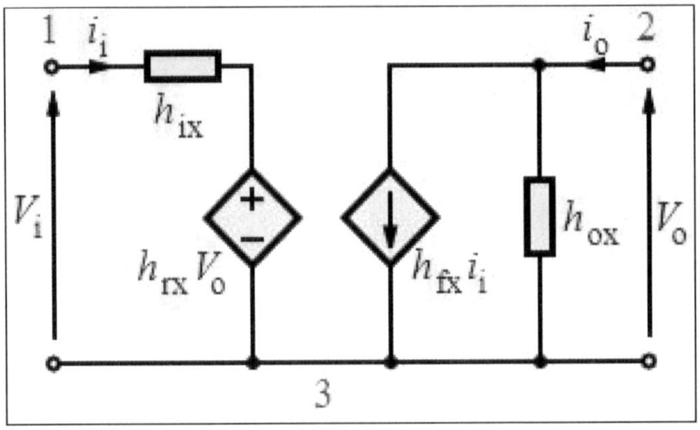

Generalized h-parameter model of an NPN BJT. *Replace x with e, b or c for CE, CB and CC topologies respectively.*

Another model commonly used to analyze BJT circuits is the *h-parameter* model, closely related to the hybrid-pi model and the y-parameter two-port, but using input current and output voltage as independent variables, rather than input and output voltages. This two-port network is particularly suited to BJTs as it lends itself easily to the analysis of circuit behaviour, and may be used to develop further accurate models. As shown, the term, *x*, in the model represents a different BJT lead depending on the topology used. For common-emitter mode the various symbols take on the specific values as:

- Terminal 1, base;
- Terminal 2, collector;
- Terminal 3 (common), emitter; giving *x* to be *e;*
- i_i, base current (i_b);
- i_o, collector current (i_c);
- V_{in}, base-to-emitter voltage (V_{BE});
- V_o, collector-to-emitter voltage (V_{CE}).

and the h-parameters are given by:

- $h_{ix} = h_{ie}$, the input impedance of the transistor (corresponding to the base resistance r_{pi}).
- $h_{rx} = h_{re}$, represents the dependence of the transistor's I_B–V_{BE} curve on the value of V_{CE}. It is usually very small and is often neglected (assumed to be zero).
- $h_{fx} = h_{fe}$, the current-gain of the transistor. This parameter is often specified as h_{FE} or the DC current-gain (β_{DC}) in datasheets.
- $h_{ox} = 1/h_{oe}$, the output impedance of transistor. The parameter h_{oe} usually corresponds to the output admittance of the bipolar transistor and has to be inverted to convert it to an impedance.

As shown, the h-parameters have lower-case subscripts and hence signify AC conditions or analyses. For DC conditions they are specified in upper-case. For the CE topology, an approximate h-parameter model is commonly used which further simplifies the circuit analysis. For this the h_{oe} and h_{re} parameters are neglected (that is, they are set to infinity and zero, respectively). The h-parameter model as shown is suited to low-frequency, small-signal analysis. For high-frequency analyses the inter-electrode capacitances that are important at high frequencies must be added.

Applications

The BJT remains a device that excels in some applications, such as discrete circuit design, due to the very wide selection of BJT types available, and because of its high transconductance and output resistance compared to MOSFETs.

The BJT is also the choice for demanding analog circuits, especially for very-high-frequency applications, such as radio-frequency circuits for wireless systems.

High-speed Digital Logic

Emitter-coupled logic (ECL) use BJTs.

Bipolar transistors can be combined with MOSFETs in an integrated circuit by using a BiCMOS process of wafer fabrication to create circuits that take advantage of the application strengths of both types of transistor.

Amplifiers

The transistor parameters α and β characterizes the current gain of the BJT. It is this gain that allows BJTs to be used as the building blocks of electronic amplifiers. The three main BJT amplifier topologies are:

- Common emitter;
- Common base;
- Common collector.

Temperature Sensors

Because of the known temperature and current dependence of the forward-biased base–emitter junction voltage, the BJT can be used to measure temperature by subtracting two voltages at two different bias currents in a known ratio.

Logarithmic Converters

Because base–emitter voltage varies as the logarithm of the base–emitter and collector–emitter currents, a BJT can also be used to compute logarithms and anti-logarithms. A diode can also perform these nonlinear functions but the transistor provides more circuit flexibility.

Field-effect Transistor

The field-effect transistor (FET) is an electronic device which uses an electric field to control the flow of current. FETs are devices with three terminals: *source*, *gate*, and *drain*. FETs control the flow of current by the application of a voltage to the gate, which in turn alters the conductivity between the drain and source.

FETs are also known as unipolar transistors since they involve single-carrier-type operation. That is, FETs use electrons or holes as charge carriers in their operation, but not both. Many different types of field effect transistors exist. Field effect transistors generally display very high input impedance at low frequencies. The most widely used field-effect transistor is the MOSFET (metal-oxide-semiconductor field-effect transistor).

FETs can be majority-charge-carrier devices, in which the current is carried predominantly by majority carriers, or minority-charge-carrier devices, in which the current is mainly due to a flow of minority carriers. The device consists of an active channel through which charge carriers, electrons or holes, flow from the source to the drain. Source and drain terminal conductors are connected to the semiconductor through ohmic contacts. The conductivity of the channel is a function of the potential applied across the gate and source terminals.

The FET's three terminals are:

1. Source (S), through which the carriers enter the channel. Conventionally, current entering the channel at S is designated by I_S.
2. Drain (D), through which the carriers leave the channel. Conventionally, current entering the channel at D is designated by I_D. Drain-to-source voltage is V_{DS}.
3. Gate (G), the terminal that modulates the channel conductivity. By applying voltage to G, one can control I_D.

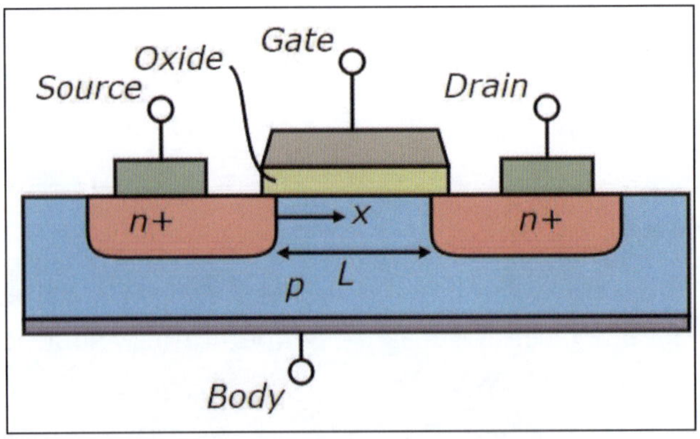

Cross section of an n-type MOSFET.

All FETs have *source*, *drain*, and *gate* terminals that correspond roughly to the *emitter*, *collector*, and *base* of BJTs. Most FETs have a fourth terminal called the *body*, *base*, *bulk*, or *substrate*. This fourth terminal serves to bias the transistor into operation; it is rare to make non-trivial use of the body terminal in circuit designs, but its presence is important when setting up the physical layout of an integrated circuit. The size of the gate, length L in the diagram, is the distance between source and drain. The *width* is the extension of the transistor, in the direction perpendicular to the cross section in the diagram (i.e., into/out of the screen). Typically the width is much larger than the length of the gate. A gate length of 1 μm limits the upper frequency to about 5 GHz, 0.2 μm to about 30 GHz.

The names of the terminals refer to their functions. The gate terminal may be thought of as controlling the opening and closing of a physical gate. This gate permits electrons to flow through or blocks their passage by creating or eliminating a channel between the source and drain. Electron-flow from the source terminal towards the drain terminal is influenced by an applied voltage. The body simply refers to the bulk of the semiconductor in which the gate, source and drain lie. Usually the body terminal is connected to the highest or lowest voltage within the circuit, depending on the type of the FET. The body terminal and the source terminal are sometimes connected together since the source is often connected to the highest or lowest voltage within the circuit, although there are several uses of FETs which do not have such a configuration, such as transmission gates and cascode circuits.

Effect of Gate Voltage on Current

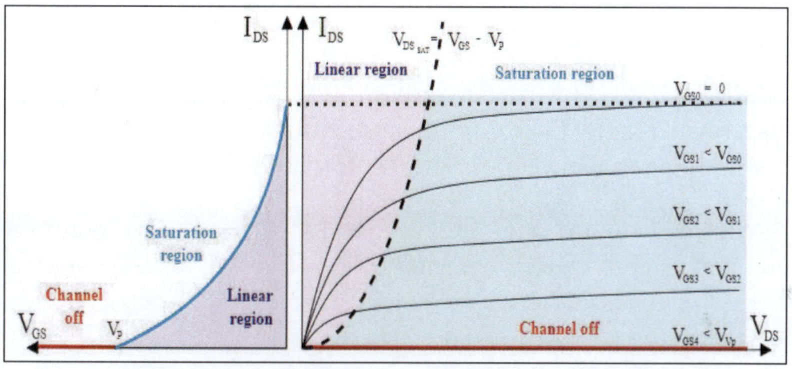

I–V characteristics and output plot of a JFET n-channel transistor.

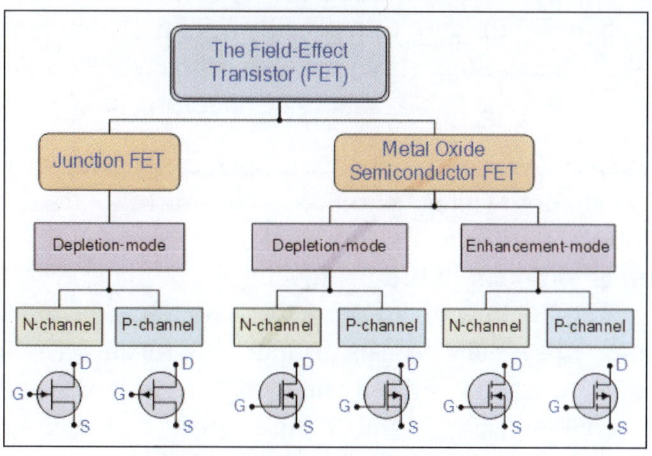

FET conventional symbol types.

The FET controls the flow of electrons (or electron holes) from the source to drain by affecting the size and shape of a "conductive channel" created and influenced by voltage (or lack of voltage) applied across the gate and source terminals. This conductive channel is the "stream" through which electrons flow from source to drain.

n-channel FET

In an n-channel "depletion-mode" device, a negative gate-to-source voltage causes a depletion region to expand in width and encroach on the channel from the sides, narrowing the channel. If the active region expands to completely close the channel, the resistance of the channel from source to drain becomes large, and the FET is effectively turned off like a switch. This is called "pinch-off", and the voltage at which it occurs is called the "pinch-off voltage". Conversely, a positive gate-to-source voltage increases the channel size and allows electrons to flow easily.

In an n-channel "enhancement-mode" device, a conductive channel does not exist naturally within the transistor, and a positive gate-to-source voltage is necessary to create one. The positive voltage attracts free-floating electrons within the body towards the gate, forming a conductive channel. But first, enough electrons must be attracted near the gate to counter the dopant ions added to the body of the FET; this forms a region with no mobile carriers called a depletion region, and the voltage at which this occurs is referred to as the threshold voltage of the FET. Further gate-to-source voltage increase will attract even more electrons towards the gate which are able to create a conductive channel from source to drain; this process is called *inversion*.

p-channel FET

In a p-channel "depletion-mode" device, a positive voltage from gate to body widens the depletion layer by forcing electrons to the gate-insulator/semiconductor interface, leaving exposed a carrier-free region of immobile, positively charged acceptor ions.

Conversely, in a p-channel "enhancement-mode" device, a conductive region does not exist and negative voltage must be used to generate a conduction channel.

Effect of Drain-to-source Voltage on Channel

For either enhancement- or depletion-mode devices, at drain-to-source voltages much less than gate-to-source voltages, changing the gate voltage will alter the channel resistance, and drain current will be proportional to drain voltage (referenced to source voltage). In this mode the FET operates like a variable resistor and the FET is said to be operating in a linear mode or ohmic mode.

If drain-to-source voltage is increased, this creates a significant asymmetrical change in the shape of the channel due to a gradient of voltage potential from source to drain. The shape of the inversion region becomes "pinched-off" near the drain end of the channel. If drain-to-source voltage is increased further, the pinch-off point of the channel begins to move away from the drain towards the source. The FET is said to be in *saturation mode*; although some authors refer to it as *active mode*, for a better analogy with bipolar transistor operating regions. The saturation mode, or the region between ohmic and saturation, is used when amplification is needed. The in-between region is sometimes considered to be part of the ohmic or linear region, even where drain current is not approximately linear with drain voltage.

Even though the conductive channel formed by gate-to-source voltage no longer connects source to drain during saturation mode, carriers are not blocked from flowing. Considering again an n-channel enhancement-mode device, a depletion region exists in the p-type body, surrounding the conductive channel and drain and source regions. The electrons which comprise the channel are free to move out of the channel through the depletion region if attracted to the drain by drain-to-source voltage. The depletion region is free of carriers and has a resistance similar to silicon. Any increase of the drain-to-source voltage will increase the distance from drain to the pinch-off point, increasing the resistance of the depletion region in proportion to the drain-to-source voltage applied. This proportional change causes the drain-to-source current to remain relatively fixed, independent of changes to the drain-to-source voltage, quite unlike its ohmic behavior in the linear mode of operation. Thus, in saturation mode, the FET behaves as a constant-current source rather than as a resistor, and can effectively be used as a voltage amplifier. In this case, the gate-to-source voltage determines the level of constant current through the channel.

Composition

FETs can be constructed from various semiconductors -- silicon is by far the most common. Most FETs are made by using conventional bulk semiconductor processing techniques, using a single crystal semiconductor wafer as the active region, or channel.

Among the more unusual body materials are amorphous silicon, polycrystalline silicon or other amorphous semiconductors in thin-film transistors or organic field-effect transistors (OFETs) that are based on organic semiconductors; often, OFET gate insulators and electrodes are made of organic materials, as well. Such FETs are manufactured using a variety of materials such as silicon carbide (SiC), gallium arsenide (GaAs), gallium nitride (GaN), and indium gallium arsenide (InGaAs).

In June 2011, IBM announced that it had successfully used graphene-based FETs in an integrated circuit. These transistors are capable of about 2.23 GHz cutoff frequency, much higher than standard silicon FETs.

Types

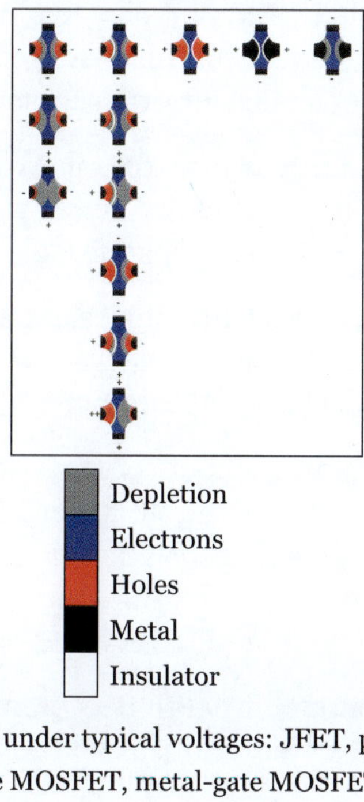

Depletion-type FETs under typical voltages: JFET, poly-silicon MOSFET, double-gate MOSFET, metal-gate MOSFET, MESFET.

The channel of a FET is doped to produce either an n-type semiconductor or a p-type semiconductor. The drain and source may be doped of opposite type to the channel, in the case of enhancement mode FETs, or doped of similar type to the channel as in depletion mode FETs. Field-effect transistors are also distinguished by the method of insulation between channel and gate. Types of FETs include:

- The JFET (junction field-effect transistor) uses a reverse biased p–n junction to separate the gate from the body.
- The MOSFET (metal–oxide–semiconductor field-effect transistor) utilizes an insulator (typically SiO_2) between the gate and the body.
- The MNOS metal–nitride–oxide–semiconductor transistor utilizes an nitride-oxide layer insulator between the gate and the body.
- The DGMOSFET (dual-gate MOSFET), a FET with two insulated gates.
- The DEPFET is a FET formed in a fully depleted substrate and acts as a sensor, amplifier and memory node at the same time. It can be used as an image (photon) sensor.
- The FREDFET (fast-reverse or fast-recovery epitaxial diode FET) is a specialized FET designed to provide a very fast recovery (turn-off) of the body diode.
- The HIGFET (heterostructure insulated-gate field-effect transistor) is now used mainly in research.
- The MODFET (modulation-doped field-effect transistor) is a high-electron-mobility transistor using a quantum well structure formed by graded doping of the active region.
- The TFET (tunnel field-effect transistor) is based on band-to-band tunneling.
- The IGBT (insulated-gate bipolar transistor) is a device for power control. It has a structure akin to a MOSFET coupled with a bipolar-like main conduction channel. These are commonly used for the 200–3000 V drain-to-source voltage range of operation. Power MOSFETs are still the device of choice for drain-to-source voltages of 1 to 200 V.
- The HEMT (high-electron-mobility transistor), also called a HFET (heterostructure FET), can be made using bandgap engineering in a ternary semiconductor such as AlGaAs. The fully depleted wide-band-gap material forms the isolation between gate and body.
- The ISFET (ion-sensitive field-effect transistor) can be used to measure ion concentrations in a solution; when the ion concentration (such as H^+, see pH electrode) changes, the current through the transistor will change accordingly.
- The BioFET (Biologically sensitive field-effect transistor) is a class of sensors/biosensors based on ISFET technology which are utilized to detect charged molecules; when a charged molecule is present, changes in the electrostatic field at the BioFET surface result in a measurable change in current through the transistor. These include EnFETs, ImmunoFETs, GenFETs, DNAFETs, CPFETs, BeetleFETs, and FETs based on ion-channels/protein binding.
- The MESFET (metal–semiconductor field-effect transistor) substitutes the p–n junction of the JFET with a Schottky barrier; and is used in GaAs and other III-V semiconductor materials.

- The NOMFET is a nanoparticle organic memory field-effect transistor.
- The GNRFET (graphene nanoribbon field-effect transistor) uses a graphene nanoribbon for its channel.
- The VeSFET (vertical-slit field-effect transistor) is a square-shaped junctionless FET with a narrow slit connecting the source and drain at opposite corners. Two gates occupy the other corners, and control the current through the slit.
- The CNTFET (carbon nanotube field-effect transistor).
- The OFET (organic field-effect transistor) uses an organic semiconductor in its channel.
- The DNAFET (DNA field-effect transistor) is a specialized FET that acts as a biosensor, by using a gate made of single-strand DNA molecules to detect matching DNA strands.
- The QFET (quantum field effect transistor) takes advantage of quantum tunneling to greatly increase the speed of transistor operation by eliminating the traditional transistor's area of electron conduction.
- The SB-FET (Schottky-barrier field-effect transistor) is a field-effect transistor with metallic source and drain contact electrodes, which create Schottky barriers at both the source-channel and drain-channel interfaces.
- The GFET is a highly sensitive graphene-based field effect transistor used as biosensors and chemical sensors. Due to the 2 dimensional structure of graphene, along with its physical properties, GFETs offer increased sensitivity, and reduced instances of 'false positives' in sensing applications.
- The Fe FET uses a ferroelectric between the gate, allowing the transistor to retain its state in the absence of bias - such devices may have application as non-volatile memory.

Advantages

The FET has high gate-to-main current resistance, on the order of 100 MΩ or more, providing a high degree of isolation between control and flow. Because base current noise will increase with shaping time, a FET typically produces less noise than a bipolar junction transistor (BJT), and is found in noise-sensitive electronics such as tuners and low-noise amplifiers for VHF and satellite receivers. It is relatively immune to radiation. It exhibits no offset voltage at zero drain current and makes an excellent signal chopper. It typically has better thermal stability than a BJT. Because they are controlled by gate charge, once the gate is closed or open, there is no additional power draw, as there would be with a bipolar junction transistor or with non-latching relays in some states. This allows extremely low-power switching, which in turn allows greater miniaturization of circuits because heat dissipation needs are reduced compared to other types of switches.

Disadvantages

A field-effect transistor has a relatively low gain–bandwidth product compared to a BJT. The MOSFET is very susceptible to overload voltages, thus requiring special handling during installation. The fragile insulating layer of the MOSFET between the gate and channel makes it vulnerable to electrostatic discharge or changes to threshold voltage during handling. This is not usually a problem after the device has been installed in a properly designed circuit.

FETs often have a very low "on" resistance and have a high "off" resistance. However, the intermediate resistances are significant, and so FETs can dissipate large amounts of power while switching. Thus efficiency can put a premium on switching quickly, but this can cause transients that can excite stray inductances and generate significant voltages that can couple to the gate and cause unintentional switching. FET circuits can therefore require very careful layout and can involve trades between switching speed and power dissipation. There is also a trade-off between voltage rating and "on" resistance, so high-voltage FETs have a relatively high "on" resistance and hence conduction losses.

Failure Modes

FETs are relatively robust, especially when operated within the temperature and electrical limitations defined by the manufacturer (proper derating). However, modern FET devices can often incorporate a body diode. If the characteristics of the body diode are not taken into consideration, the FET can experience slow body diode behavior, where a parasitic transistor will turn on and allow high current to be drawn from drain to source when the FET is off.

Uses

The most commonly used FET is the MOSFET. The CMOS (complementary metal oxide semiconductor) process technology is the basis for modern digital integrated circuits. This process technology uses an arrangement where the (usually "enhancement-mode") p-channel MOSFET and n-channel MOSFET are connected in series such that when one is on, the other is off.

In FETs, electrons can flow in either direction through the channel when operated in the linear mode. The naming convention of drain terminal and source terminal is somewhat arbitrary, as the devices are typically (but not always) built symmetrical from source to drain. This makes FETs suitable for switching analog signals between paths (multiplexing). With this concept, one can construct a solid-state mixing board, for example. FET is commonly used as an amplifier. For example, due to its large input resistance and low output resistance, it is effective as a buffer in common-drain (source follower) configuration.

IGBTs are used in switching internal combustion engine ignition coils, where fast switching and voltage blocking capabilities are important.

Source-gated Transistor

Source-gated transistors are more robust to manufacturing and environmental issues in large-area electronics such as display screens, but are slower in operation than FETs.

MOSFET

The metal–oxide–semiconductor field-effect transistor (MOSFET, MOS-FET, or MOS FET), also known as the metal–oxide–silicon transistor (MOS transistor, or MOS), is a type of field-effect transistor that is fabricated by the controlled oxidation of a semiconductor, typically silicon. It has an insulated gate, whose voltage determines the conductivity of the device. This ability to change conductivity with the amount of applied voltage can be used for amplifying or switching electronic signals. The MOSFET is the basic element in most modern electronic equipment. Since its invention by Mohamed Atalla and Dawon Kahng at Bell Labs in November 1959, the MOSFET has become the most widely manufactured device in history, with an estimated total of 13 sextillion MOS transistors manufactured as of 2018.

A key advantage of a MOSFET is that it requires almost no input current to control the load current, when compared with bipolar transistors (bipolar junction transistors, or BJTs). In an *enhancement mode* MOSFET, voltage applied to the gate terminal increases the conductivity of the device. In *depletion mode* transistors, voltage applied at the gate reduces the conductivity. MOSFETs are capable of high scalability (Moore's law and Dennard scaling), with increasing miniaturisation, and can be easily scaled down to smaller dimensions. They also consume much less power, and allow higher density, than bipolar transistors. The MOSFET is also cheaper and has relatively simple processing steps, resulting in a high manufacturing yield. Since MOSFETs can be made with either p-type or n-type semiconductors (PMOS or NMOS logic, respectively), complementary pairs of MOS transistors can be used to make switching circuits with very low power consumption, in the form of CMOS (complementary MOS) logic.

The name "metal–oxide–semiconductor" (MOS) typically refers to a metal gate, oxide insulation, and semiconductor (typically silicon). However, the "metal" in the name MOSFET is sometimes a misnomer, because the gate material can also be a layer of polysilicon (polycrystalline silicon). Similarly, "oxide" in the name can also be a misnomer, as different dielectric materials can be used with the aim of obtaining strong channels with smaller applied voltages.

The MOSFET is by far the most widely used transistor in both digital circuits and analog circuits, and is the fundamental building block of digital electronics. It revolutionised the electronics industry, and has been central to the microcomputer revolution, digital revolution, information revolution, and information age. MOSFETs are fundamental to high-density integrated circuits such as memory chips and microprocessors, each using thousands to billions of MOS transistors. Discrete MOSFETs are also the most common power devices, and are used in applications such as power supplies, RF amplifiers, audio amplifiers, automotive electronics, the Internet, wireless mobile networks, LCD displays, X-ray detectors, spacecraft, and satellites, among other uses. The US Patent and Trademark Office calls the MOSFET a "groundbreaking invention that transformed life and culture around the world".

Importance

The MOSFET revolutionized the electronics industry, including power electronics, electronic signal processing, control systems, and computers. The MOSFET was central to the microelectronics revolution and the microcomputer revolution in the late 20th century. The MOS transistor has since been the fundamental building block of modern digital electronics during the digital revolution, information revolution, and 21st-century information age. MOSFETs have been the driving force behind the computer revolution, and the technologies enabled by it, such as computers, the internet, and smartphones. The rapid progress of the electronics industry during the late 20th to early 21st centuries was achieved by rapid MOSFET scaling (Dennard scaling and Moore's law), down to the level of nanoelectronics in the early 21st century.

The MOSFET is the basic element in most modern electronic equipment. It is the most widely used semiconductor device in the world, and the most common transistor in electronics and communications technology. It has been described as the "workhorse of the electronics industry" due to being the building block of every microprocessor, memory chip and telecommunication circuit in use, and the basis for nearly every silicon integrated circuit (IC), TFT LCD display, digital wireless telecommunication network, and graphics processing unit (GPU), among other uses.

The MOSFET is the most widely manufactured device in history. As of 2013, billions of MOS transistors are manufactured every day. As of 2018, an estimated total of 13 sextillion MOS transistors have been manufactured. Digital integrated circuits such as microprocessors and memory devices contain thousands to billions of integrated MOSFET transistors on each device, providing the basic switching functions required to implement logic gates and data storage. There are also memory devices which contain at least a trillion MOS transistors, such as a 256 GB microSD memory card, larger than the number of stars in the Milky Way galaxy.

Power MOSFETs and MOS integrated circuits are fundamental to the Internet, and the communications infrastructure that enables the Internet. MOSFETs are fundamental to digital wireless telecommunication networks, such as mobile networks. Most of the essential elements of mobile networks use MOSFETs, including the handheld mobile devices themselves, along with the mobile tranceivers, base station modules, routers, and RF power amplifiers. The thin-film transistor (TFT), a type of MOSFET, is the basis for TFT LCD technology, used by liquid-crystal display (LCD) displays and LCD televisions. TFTs are widely used in active-matrix LCD (AM LCD) and OLED (organic light-emitting diode) displays, as well as X-ray detectors.

The US Patent and Trademark Office calls the MOSFET a "groundbreaking invention that transformed life and culture around the world" and the Computer History Museum credits it with "irrevocably changing the human experience" as "the base technology" of the late 20th to early 21st centuries. The MOSFET is included on the list of IEEE milestones in electronics, and its inventors Mohamed Atalla and Dawon Kahng entered the National Inventors Hall of Fame in 2009.

Composition

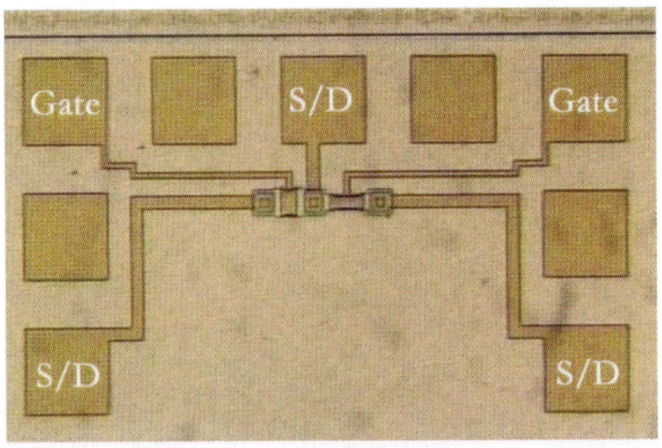

Photomicrograph of two metal-gate MOSFETs in a test pattern. Probe pads for two gates and three source/drain nodes are labeled.

Usually the semiconductor of choice is silicon. Recently, some chip manufacturers, most notably IBM and Intel, have started using a chemical compound of silicon and germanium (SiGe) in MOSFET channels. Unfortunately, many semiconductors with better electrical properties than silicon, such as gallium arsenide, do not form good semiconductor-to-insulator interfaces, and thus are not suitable for MOSFETs. Research continues on creating insulators with acceptable electrical characteristics on other semiconductor materials.

To overcome the increase in power consumption due to gate current leakage, a high-κ dielectric is used instead of silicon dioxide for the gate insulator, while polysilicon is replaced by metal gates (e.g. Intel, 2009).

The gate is separated from the channel by a thin insulating layer, traditionally of silicon dioxide and later of silicon oxynitride. Some companies have started to introduce a high-κ dielectric and metal gate combination in the 45 nanometer node.

When a voltage is applied between the gate and body terminals, the electric field generated penetrates through the oxide and creates an *inversion layer* or *channel* at the semiconductor-insulator interface. The inversion layer provides a channel through which current can pass between source and drain terminals. Varying the voltage between the gate and body modulates the conductivity of this layer and thereby controls the current flow between drain and source. This is known as enhancement mode.

Operation

Metal–oxide–semiconductor Structure

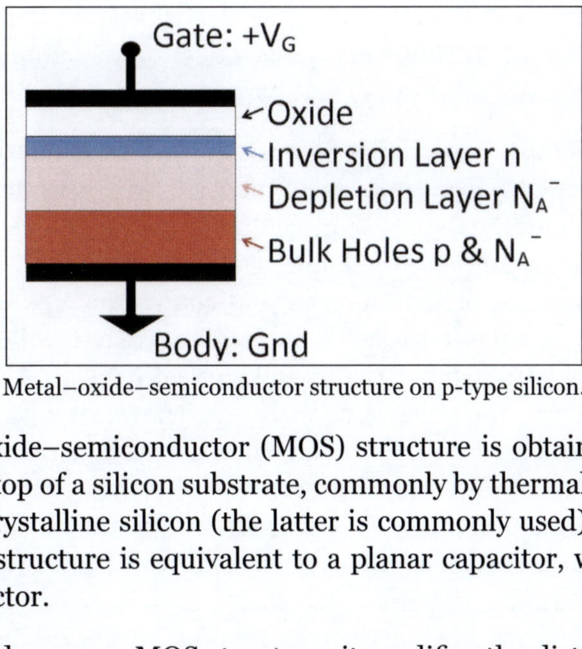

Metal–oxide–semiconductor structure on p-type silicon.

The traditional metal–oxide–semiconductor (MOS) structure is obtained by growing a layer of silicon dioxide (SiO_2) on top of a silicon substrate, commonly by thermal oxidation and depositing a layer of metal or polycrystalline silicon (the latter is commonly used). As the silicon dioxide is a dielectric material, its structure is equivalent to a planar capacitor, with one of the electrodes replaced by a semiconductor.

When a voltage is applied across a MOS structure, it modifies the distribution of charges in the semiconductor. If we consider a p-type semiconductor (with N_A the density of acceptors, p the density of holes; $p = N_A$ in neutral bulk), a positive voltage, V_{GB}, from gate to body creates a depletion layer by forcing the positively charged holes away from the gate-insulator/semiconductor interface, leaving exposed a carrier-free region of immobile, negatively charged acceptor ions. If V_{GB} is high enough, a high concentration of negative charge carriers forms in an *inversion layer* located in a thin layer next to the interface between the semiconductor and the insulator.

Conventionally, the gate voltage at which the volume density of electrons in the inversion layer is the same as the volume density of holes in the body is called the threshold voltage. When the voltage between transistor gate and source (V_{GS}) exceeds the threshold voltage (V_{th}), the difference is known as overdrive voltage.

This structure with p-type body is the basis of the n-type MOSFET, which requires the addition of n-type source and drain regions.

MOS Capacitors and Band Diagrams

The MOS capacitor structure is the heart of the MOSFET. Consider a MOS capacitor where the silicon base is of p-type. If a positive voltage is applied at the gate, holes which are at the surface of the p-type substrate will be repelled by the electric field generated by the voltage applied. At first, the holes will simply be repelled and what will remain on the surface will be immobile (negative) atoms of the acceptor type, which creates a depletion region on the surface. Remember that a hole is created by an acceptor atom, e.g. Boron, which has one less electron than Silicon. One might ask how can holes be repelled if they are actually non-entities? The answer is that what really happens is not that a hole is repelled, but electrons are attracted by the positive field, and fill these holes, creating a depletion region where no charge carriers exist because the electron is now fixed onto the atom and immobile.

As the voltage at the gate increases, there will be a point at which the surface above the depletion region will be converted from p-type into n-type, as electrons from the bulk area will start to get attracted by the larger electric field. This is known as *inversion*. The threshold voltage at which this conversion happens is one of the most important parameters in a MOSFET.

In the case of a p-type bulk, inversion happens when the intrinsic energy level at the surface becomes smaller than the Fermi level at the surface. One can see this from a band diagram. Remember that the Fermi level defines the type of semiconductor in discussion. If the Fermi level is equal to the Intrinsic level, the semiconductor is of intrinsic, or pure type. If the Fermi level lies closer to the conduction band (valence band) then the semiconductor type will be of n-type (p-type). Therefore, when the gate voltage is increased in a positive sense (for the given example), this will "bend" the intrinsic energy level band so that it will curve downwards towards the valence band. If the Fermi level lies closer to the valence band (for p-type), there will be a point when the Intrinsic level will start to cross the Fermi level and when the voltage reaches the threshold voltage, the intrinsic level does cross the Fermi level, and that is what is known as inversion. At that point, the surface of the semiconductor is inverted from p-type into n-type. If the Fermi level lies above the Intrinsic level, the semiconductor is of n-type, therefore at Inversion, when the Intrinsic level reaches and crosses the Fermi level (which lies closer to the valence band), the semiconductor type changes at the surface as dictated by the relative positions of the Fermi and Intrinsic energy levels.

Structure and Channel Formation

A MOSFET is based on the modulation of charge concentration by a MOS capacitance between a *body* electrode and a *gate* electrode located above the body and insulated from all other device regions by a gate dielectric layer. If dielectrics other than an oxide are employed, the device may be referred to as a metal-insulator-semiconductor FET (MISFET). Compared to the MOS capacitor, the MOSFET includes two additional terminals (*source* and *drain*), each connected to individual highly doped regions that are separated by the body region. These regions can be either p or n type, but they must both be of the same type, and of opposite type to the body region. The source and drain (unlike the body) are highly doped as signified by a "+" sign after the type of doping.

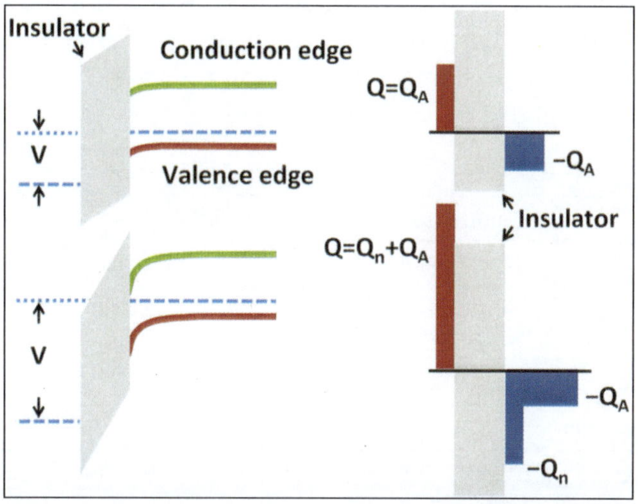

Channel formation in nMOS MOSFET shown as band diagram: Top panels: An applied gate voltage bends bands, depleting holes from surface (left). The charge inducing the bending is balanced by a layer of negative acceptor-ion charge (right). Bottom panel: A larger applied voltage further depletes holes but conduction band lowers enough in energy to populate a conducting channel.

If the MOSFET is an n-channel or nMOS FET, then the source and drain are *n+* regions and the body is a *p* region. If the MOSFET is a p-channel or pMOS FET, then the source and drain are *p+* regions and the body is a *n* region. The source is so named because it is the source of the charge carriers (electrons for n-channel, holes for p-channel) that flow through the channel; similarly, the drain is where the charge carriers leave the channel.

The occupancy of the energy bands in a semiconductor is set by the position of the Fermi level relative to the semiconductor energy-band edges.

With sufficient gate voltage, the valence band edge is driven far from the Fermi level, and holes from the body are driven away from the gate.

At larger gate bias still, near the semiconductor surface the conduction band edge is brought close to the Fermi level, populating the surface with electrons in an *inversion layer* or *n-channel* at the interface between the p region and the oxide. This conducting channel extends between the source and the drain, and current is conducted through it when a voltage is applied between the two electrodes. Increasing the voltage on the gate leads to a higher electron density in the inversion layer and therefore increases the current flow between the source and drain. For gate voltages below the threshold value, the channel is lightly populated, and only a very small subthreshold leakage current can flow between the source and the drain.

When a negative gate-source voltage (positive source-gate) is applied, it creates a *p-channel* at the surface of the n region, analogous to the n-channel case, but with opposite polarities of charges and voltages. When a voltage less negative than the threshold value (a negative voltage for the p-channel) is applied between gate and source, the channel disappears and only a very small subthreshold current can flow between the source and the drain. The device may comprise a silicon on insulator device in which a buried oxide is formed below a thin semiconductor layer. If the channel region between the gate dielectric and the buried oxide region is very thin, the channel is referred to as an ultrathin channel region with the source and drain regions formed on either side in or above the thin semiconductor layer. Other semiconductor materials may be employed. When the source and drain regions are formed above the channel in whole or in part, they are referred to as raised source/drain regions.

Comparison of n- and p-type MOSFETs

Parameter		nMOSFET	pMOSFET
Source/drain type		n-type	p-type
Channel type (MOS capacitor)		n-type	p-type
Gate type	Polysilicon	n+	p+
	Metal	φ_m ~ Si conduction band	φ_m ~ Si valence band
Well type		p-type	n-type
Threshold voltage, V_{th}		• Positive (enhancement) • Negative (depletion)	• Negative (enhancement) • Positive (depletion)
Band-bending		Downwards	Upwards
Inversion layer carriers		Electrons	Holes
Substrate type		p-type	n-type

Modes of Operation

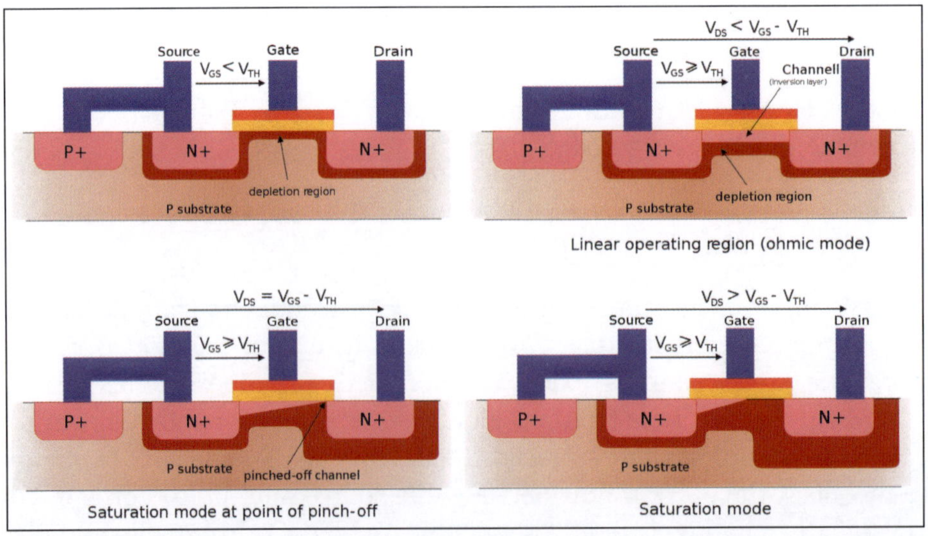

Source tied to the body to ensure no body bias: top left: Subthreshold, top right: Ohmic mode, bottom left: Active mode at onset of pinch-off, bottom right: Active mode well into pinch-off – channel length modulation evident.

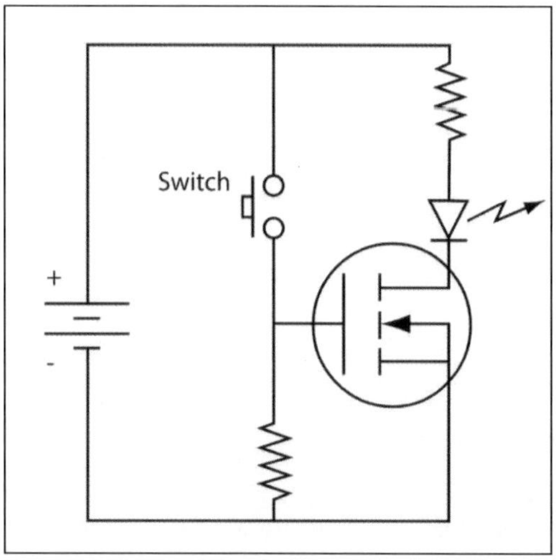

Example application of an n-channel MOSFET. When the switch is pushed, the LED lights up.

The operation of a MOSFET can be separated into three different modes, depending on the voltages at the terminals.

For an *enhancement-mode, n-channel MOSFET*, the three operational modes are:

Cutoff, subthreshold, and weak-inversion mode,

When $V_{GS} < V_{th}$:

where V_{GS} is gate-to-source bias and V_{th} is the threshold voltage of the device.

According to the basic threshold model, the transistor is turned off, and there is no conduction between drain and source. A more accurate model considers the effect of thermal energy on the Fermi–Dirac distribution of electron energies which allow some of the more energetic electrons at the source to enter the channel and flow to the drain. This results in a subthreshold current that is an exponential function of gate-source voltage. While the current between drain and source should ideally be zero when the transistor is being used as a turned-off switch, there is a weak-inversion current, sometimes called subthreshold leakage.

In weak inversion where the source is tied to bulk, the current varies exponentially with V_{GS} as given approximately by:

$$I_D \approx I_{D0} e^{\frac{V_{GS}-V_{th}}{nV_T}},$$

where I_{D0} = current at $V_{GS} = V_{th}$, the thermal voltage $V_T = kT/q$ and the slope factor n is given by:

$$n = 1 + \frac{C_{dep}}{C_{ox}},$$

with C_{dep} = capacitance of the depletion layer and C_{ox} = capacitance of the oxide layer. This equation is generally used, but is only an adequate approximation for the source tied to the bulk. For the source not tied to the bulk, the subthreshold equation for drain current in saturation is,

$$I_D \approx I_{D0} e^{\frac{\kappa(V_G - V_{th}) - V_S}{V_T}},$$

where the κ is the channel divider that is given by:

$$\kappa = \frac{C_{ox}}{C_{ox} + C_D},$$

with C_D = capacitance of the depletion layer and C_{ox} = capacitance of the oxide layer. In a long-channel device, there is no drain voltage dependence of the current once $V_{DS} \gg V_T$, but as channel length is reduced drain-induced barrier lowering introduces drain voltage dependence that depends in a complex way upon the device geometry (for example, the channel doping, the junction doping and so on). Frequently, threshold voltage V_{th} for this mode is defined as the gate voltage at which a selected value of current I_{Do} occurs, for example, I_{Do} = 1 μA, which may not be the same V_{th}-value used in the equations for the following modes.

Some micropower analog circuits are designed to take advantage of subthreshold conduction. By working in the weak-inversion region, the MOSFETs in these circuits deliver the highest possible transconductance-to-current ratio, namely: $g_m / I_D = 1/(nV_T)$, almost that of a bipolar transistor.

The subthreshold *I–V curve* depends exponentially upon threshold voltage, introducing a strong dependence on any manufacturing variation that affects threshold voltage; for example: variations in oxide thickness, junction depth, or body doping that change the degree of drain-induced barrier lowering. The resulting sensitivity to fabricational variations complicates optimization for leakage and performance.

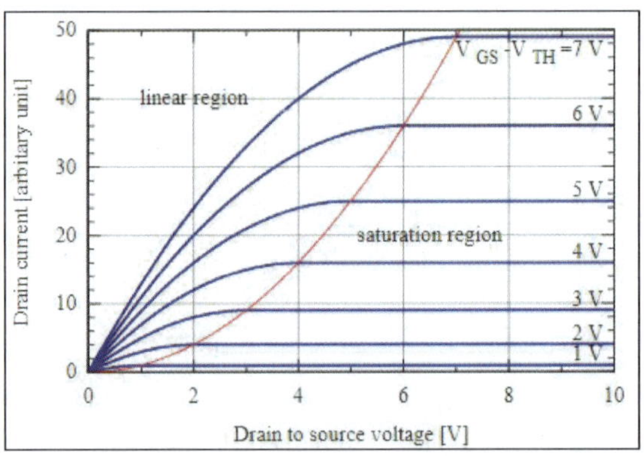

MOSFET drain current vs. drain-to-source voltage for several values of ; the boundary between *linear* (*Ohmic*) and *saturation* (*active*) modes is indicated by the upward curving parabola.

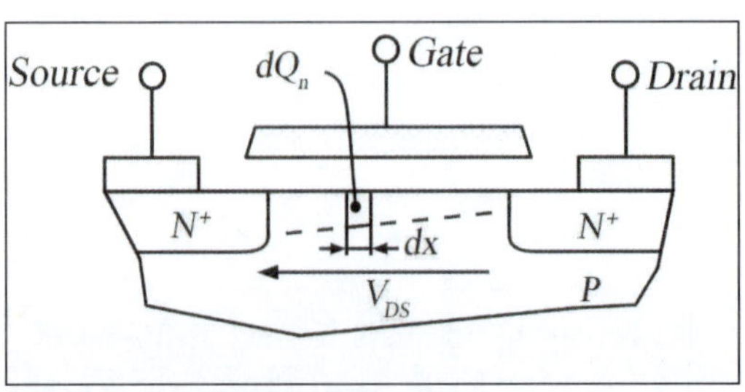

Cross section of a MOSFET operating in the linear (Ohmic) region; strong inversion region present even near drain.

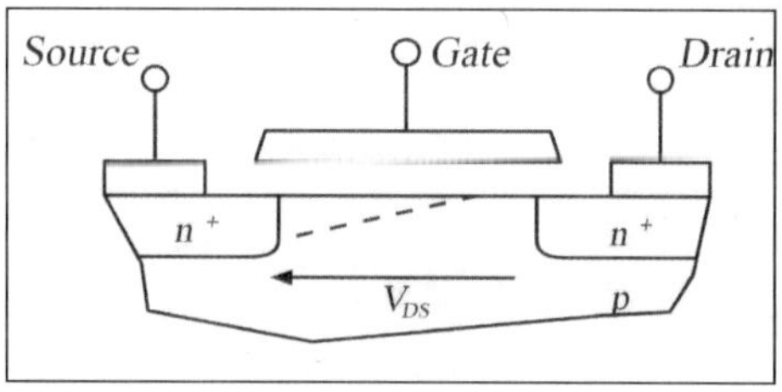

Cross section of a MOSFET operating in the saturation (active) region; channel exhibits channel pinching near drain.

Triode mode or linear region (also known as the ohmic mode),

When $V_{GS} > V_{th}$ and $V_{DS} < V_{GS} - V_{th}$:

The transistor is turned on, and a channel has been created which allows current between the drain and the source. The MOSFET operates like a resistor, controlled by the gate voltage relative to both the source and drain voltages. The current from drain to source is modeled as:

$$I_D = \mu_n C_{ox} \frac{W}{L}\left((V_{GS} - V_{th}) V_{DS} - \frac{V_{DS}^2}{2} \right)$$

where μ_n is the charge-carrier effective mobility, W is the gate width, L is the gate length and C_{ox} is the gate oxide capacitance per unit area. The transition from the exponential subthreshold region to the triode region is not as sharp as the equations suggest.

Saturation or Active Mode

When $V_{GS} > V_{th}$ and $V_{DS} \geq (V_{GS} - V_{th})$:

The switch is turned on, and a channel has been created, which allows current between the drain and source. Since the drain voltage is higher than the source voltage, the electrons spread out, and conduction is not through a narrow channel but through a broader, two- or three-dimensional current distribution extending away from the interface and deeper in the substrate. The onset of this region is also known as pinch-off to indicate the lack of channel region near the drain. Although the channel does not extend the full length of the device, the electric field between the drain and the channel is very high, and conduction continues. The drain current is now weakly dependent upon drain voltage and controlled primarily by the gate-source voltage, and modeled approximately as:

$$I_D = \frac{\mu_n C_{ox}}{2} \frac{W}{L} [V_{GS} - V_{th}]^2 [1 + \lambda (V_{DS} - V_{DSsat})].$$

The additional factor involving λ, the channel-length modulation parameter, models current dependence on drain voltage due to the Early effect, or channel length modulation. According to this equation, a key design parameter, the MOSFET transconductance is:

$$g_m = \frac{\partial I_D}{\partial V_{GS}} = \frac{2 I_D}{V_{GS} - V_{th}} = \frac{2 I_D}{V_{ov}},$$

where the combination $V_{ov} = V_{GS} - V_{th}$ is called the overdrive voltage, and where $V_{DSsat} = V_{GS} - V_{th}$ accounts for a small discontinuity in I_D which would otherwise appear at the transition between the triode and saturation regions.

Another key design parameter is the MOSFET output resistance r_{out} given by:

$$r_{out} = \frac{1}{\lambda I_D}.$$

r_{out} is the inverse of g_{DS} where $g_{DS} = \frac{\partial I_{DS}}{\partial V_{DS}}$. I_D is the expression in saturation region.

If λ is taken as zero, an infinite output resistance of the device results that leads to unrealistic circuit predictions, particularly in analog circuits.

As the channel length becomes very short, these equations become quite inaccurate. New physical effects arise. For example, carrier transport in the active mode may become limited by velocity saturation. When velocity saturation dominates, the saturation drain current is more nearly linear than quadratic in V_{GS}. At even shorter lengths, carriers transport with near zero scattering, known as quasi-ballistic transport. In the ballistic regime, the carriers travel at an injection velocity that may exceed the saturation velocity and approaches the Fermi velocity at high inversion charge density. In addition, drain-induced barrier lowering increases off-state (cutoff) current and requires an increase in threshold voltage to compensate, which in turn reduces the saturation current.

Body Effect

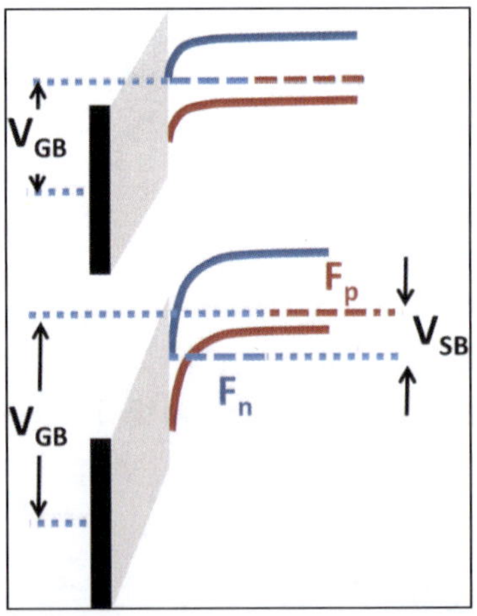

Band diagram showing body effect. V_{SB} splits Fermi levels F_n for electrons and F_p for holes, requiring larger V_{GB} to populate the conduction band in an nMOS MOSFET.

The occupancy of the energy bands in a semiconductor is set by the position of the Fermi level relative to the semiconductor energy-band edges. Application of a source-to-substrate reverse bias of the source-body pn-junction introduces a split between the Fermi levels for electrons and holes, moving the Fermi level for the channel further from the band edge, lowering the occupancy of the channel. The effect is to increase the gate voltage necessary to establish the channel, as seen in the figure. This change in channel strength by application of reverse bias is called the 'body effect'.

Simply put, using an nMOS example, the gate-to-body bias V_{GB} positions the conduction-band energy levels, while the source-to-body bias V_{SB} positions the electron Fermi level near the interface, deciding occupancy of these levels near the interface, and hence the strength of the inversion layer or channel.

The body effect upon the channel can be described using a modification of the threshold voltage, approximated by the following equation:

$$V_{TB} = V_{T0} + \gamma\left(\sqrt{V_{SB} + 2\varphi_B} - \sqrt{2\varphi_B}\right),$$

where V_{TB} is the threshold voltage with substrate bias present, and V_{T0} is the zero-V_{SB} value of threshold voltage, γ is the body effect parameter, and $2\varphi_B$ is the approximate potential drop between surface and bulk across the depletion layer when $V_{SB} = 0$ and gate bias is sufficient to ensure that a channel is present. As this equation shows, a reverse bias $V_{SB} > 0$ causes an increase in threshold voltage V_{TB} and therefore demands a larger gate voltage before the channel populates.

The body can be operated as a second gate, and is sometimes referred to as the "back gate"; the body effect is sometimes called the "back-gate effect".

Circuit Symbols

A variety of symbols are used for the MOSFET. The basic design is generally a line for the channel with the source and drain leaving it at right angles and then bending back at right angles into the same direction as the channel. Sometimes three line segments are used for enhancement mode and a solid line for depletion mode. Another line is drawn parallel to the channel for the gate.

The *bulk* or *body* connection, if shown, is shown connected to the back of the channel with an arrow indicating pMOS or nMOS. Arrows always point from P to N, so an NMOS (N-channel in P-well or P-substrate) has the arrow pointing in (from the bulk to the channel). If the bulk is connected to the source (as is generally the case with discrete devices) it is sometimes angled to meet up with the source leaving the transistor. If the bulk is not shown (as is often the case in IC design as they are generally common bulk) an inversion symbol is sometimes used to indicate PMOS, alternatively an arrow on the source may be used in the same way as for bipolar transistors (out for nMOS, in for pMOS).

Comparison of enhancement-mode and depletion-mode MOSFET symbols, along with JFET symbols.

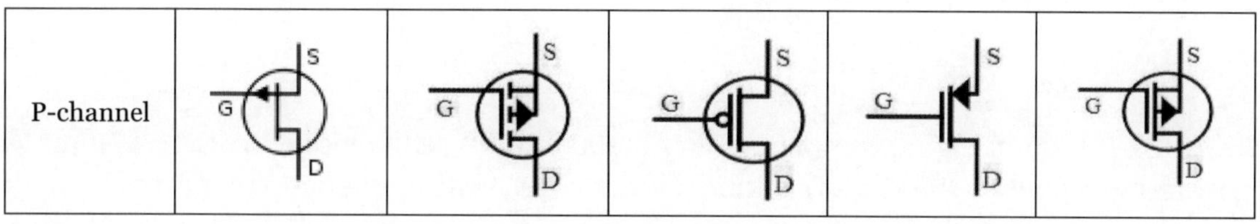

N-channel	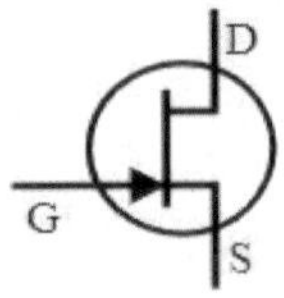	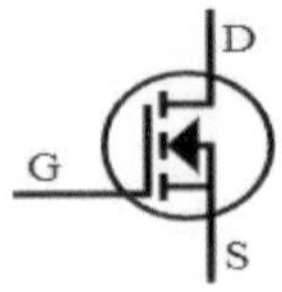	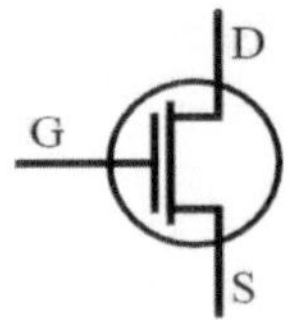	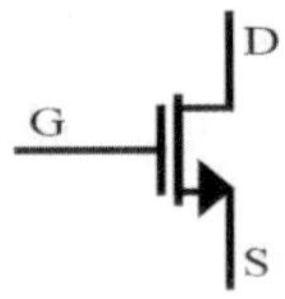	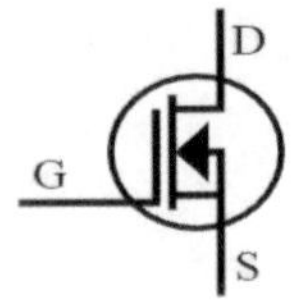
	JFET	MOSFET enh.	MOSFET enh. (no bulk)		MOSFET dep.

In schematics where G, S, D are not labeled, the detailed features of the symbol indicate which terminal is source and which is drain. For enhancement-mode and depletion-mode MOSFET symbols (in columns two and five), the source terminal is the one connected to the triangle. Additionally, in this diagram, the gate is shown as an "L" shape, whose input leg is closer to S than D, also indicating which is which. However, these symbols are often drawn with a "T" shaped gate, so it is the triangle which must be relied upon to indicate the source terminal.

For the symbols in which the bulk, or body, terminal is shown, it is here shown internally connected to the source (i.e., the black triangles in the diagrams in columns 2 and 5). This is a typical configuration, but by no means the only important configuration. In general, the MOSFET is a four-terminal device, and in integrated circuits many of the MOSFETs share a body connection, not necessarily connected to the source terminals of all the transistors.

Types of MOSFET

PMOS and NMOS Logic

P-type MOS (PMOS) logic uses p-channel MOSFETs to implement logic gates and other digital circuits. N-type MOS (NMOS) logic uses n-channel MOSFETs to implement logic gates and other digital circuits.

For devices of equal current driving capability, n-channel MOSFETs can be made smaller than p-channel MOSFETs, due to p-channel charge carriers (holes) having lower mobility than do n-channel charge carriers (electrons), and producing only one type of MOSFET on a silicon substrate is cheaper and technically simpler. These were the driving principles in the design of NMOS logic which uses n-channel MOSFETs exclusively. However, neglecting leakage current, unlike CMOS logic, NMOS logic consumes power even when no switching is taking place.

Mohamed Atalla and Dawon Kahng, after they invented the MOSFET, fabricated both pMOS and nMOS devices with a 20 µm process in 1960. However, the nMOS devices were impractical, and only the pMOS type were practical working devices. A more practical NMOS process was developed several years later. NMOS was initially faster than CMOS, thus NMOS was more widely used for computers in the 1970s. With advances in technology, CMOS logic displaced NMOS logic in the mid-1980s to become the preferred process for digital chips.

Complementary MOS (CMOS)

The MOSFET is used in digital complementary metal–oxide–semiconductor (CMOS) logic, which uses p- and n-channel MOSFETs as building blocks. Overheating is a major concern in integrated circuits since ever more transistors are packed into ever smaller chips. CMOS logic reduces power consumption because no current flows (ideally), and thus no power is consumed, except when the inputs to logic gates are being switched. CMOS accomplishes this current reduction by complementing every nMOSFET with a pMOSFET and connecting both gates and both drains together. A high voltage on the gates will cause the nMOSFET to conduct and the pMOSFET not to conduct and a low voltage on the gates causes the reverse. During the switching time as the voltage goes from one state to another, both MOSFETs will conduct briefly. This arrangement greatly reduces power consumption and heat generation.

CMOS was developed by Chih-Tang Sah and Frank Wanlass at Fairchild Semiconductor in 1963. CMOS had lower power consumption, but was initially slower than NMOS, which was more widely used for computers in the 1970s. In 1978, Hitachi introduced the twin-well CMOS process, which allowed CMOS to match the performance of NMOS with less power consumption. The twin-well CMOS process eventually overtook NMOS as the most common semiconductor manufacturing process for computers in the 1980s.

Floating-gate MOSFET (FGMOS)

The floating-gate MOSFET (FGMOS) is a type of MOSFET where the gate is electrically isolated, creating a floating node in DC, and a number of secondary gates or inputs are deposited above the floating gate (FG) and are electrically isolated from it. The first report of a floating-gate MOSFET (FGMOS) was made by Dawon Kahng (co-inventor of the original MOSFET) and Simon Sze in 1967.

The FGMOS is commonly used as a floating-gate memory cell, the digital storage element in EPROM, EEPROM and flash memories. Other uses of the FGMOS include a neuronal computational element in neural networks, analog storage element, digital potentiometers and single-transistor DACs.

Depletion-mode

There are *depletion-mode* MOSFET devices, which are less commonly used than the standard *enhancement-mode* devices already described. These are MOSFET devices that are doped so that a channel exists even with zero voltage from gate to source. To control the channel, a negative voltage is applied to the gate (for an n-channel device), depleting the channel, which reduces the current flow through the device. In essence, the depletion-mode device is equivalent to a normally closed (on) switch, while the enhancement-mode device is equivalent to a normally open (off) switch.

Due to their low noise figure in the RF region, and better gain, these devices are often preferred to bipolars in RF front-ends such as in TV sets.

Depletion-mode MOSFET families include BF960 by Siemens and Telefunken, and the BF980 in the 1980s by Philips (later to become NXP Semiconductors), whose derivatives are still used in AGC and RF mixer front-ends.

Metal-insulator-semiconductor Field-effect Transistor (MISFET)

Metal-insulator-semiconductor field-effect-transistor, or *MISFET*, is a more general term than *MOSFET* and a synonym to *insulated-gate field-effect transistor* (IGFET). All MOSFETs are MISFETs, but not all MISFETs are MOSFETs.

The gate dielectric insulator in a MISFET is silicon dioxide in a MOSFET, but other materials can also be employed. The gate dielectric lies directly below the gate electrode and above the channel of the MISFET. The term *metal* is historically used for the gate material, even though now it is usually highly doped polysilicon or some other non-metal.

Insulator types may be:

- Silicon dioxide, in MOSFETs;
- Organic insulators (e.g., undoped trans-polyacetylene; cyanoethyl pullulan, CEP), for organic-based FETs.

Thin-film Transistor (TFT)

The thin-film transistor (TFT) is a type of MOSFET distinct from the standard bulk MOSFET. The first TFT was invented by Paul K. Weimer at RCA in 1962, building on the earlier work of Atalla and Kahng on MOSFETs.

Power MOSFET

Power MOSFETs have a different structure. As with most power devices, the structure is vertical and not planar. Using a vertical structure, it is possible for the transistor to sustain both high blocking voltage and high current. The voltage rating of the transistor is a function of the doping and thickness of the N-epitaxial layer, while the current rating is a function of the channel width (the wider the channel, the higher the current). In a planar structure, the current and breakdown voltage ratings are both a function of the channel dimensions (respectively width and length of the channel), resulting in inefficient use of the "silicon estate". With the vertical structure, the component area is roughly proportional to the current it can sustain, and the component thickness (actually the N-epitaxial layer thickness) is proportional to the breakdown voltage.

Two power MOSFETs in D2PAK surface-mount packages. Operating as switches, each of these components can sustain a blocking voltage of 120 V in the *off* state, and can conduct a continuous current of 30 A in the *on* state, dissipating up to about 100 W and controlling a load of over 2000 W. A matchstick is pictured for scale.

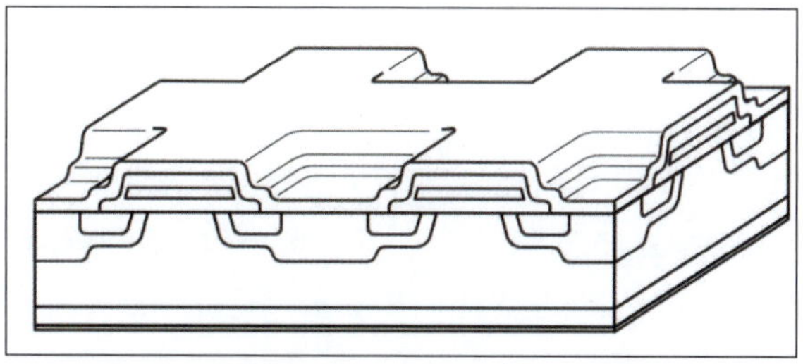

Cross section of a power MOSFET, with square cells. A typical transistor is constituted of several thousand cells.

Power MOSFETs with lateral structure are mainly used in high-end audio amplifiers and high-power PA systems. Their advantage is a better behaviour in the saturated region (corresponding to the linear region of a bipolar transistor) than the vertical MOSFETs. Vertical MOSFETs are designed for switching applications.

The power MOSFET, which is commonly used in power electronics, was developed in the early 1970s. The power MOSFET enables low gate drive power, fast switching speed, and advanced paralleling capability.

Double-diffused Metal–Oxide–Semiconductor (DMOS)

There are *LDMOS* (lateral double-diffused metal oxide semiconductor) and *VDMOS* (vertical double-diffused metal oxide semiconductor). Most power MOSFETs are made using this technology.

Multi-gate Field-effect Transistor (MuGFET)

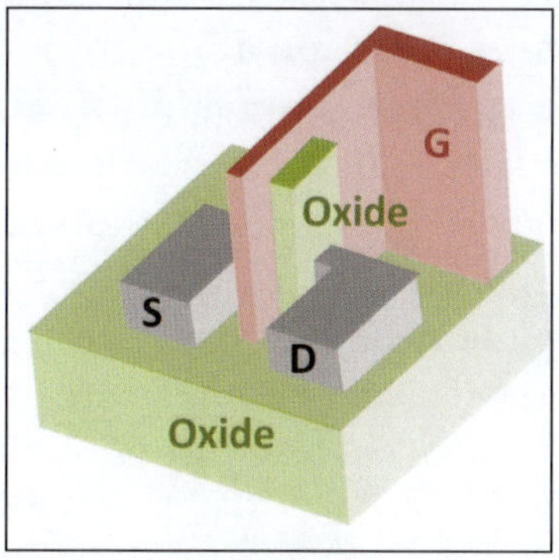

A FinFET.

The dual-gate MOSFET has a tetrode configuration, where both gates control the current in the device. It is commonly used for small-signal devices in radio frequency applications where biasing the drain-side gate at constant potential reduces the gain loss caused by Miller effect, replacing two separate transistors in cascode configuration. Other common uses in RF circuits include gain control and mixing (frequency conversion). The *tetrode* description, though accurate, does not replicate the vacuum-tube tetrode. Vacuum-tube tetrodes, using a screen grid, exhibit much lower grid-plate capacitance and much higher output impedance and voltage gains than triode vacuum tubes. These improvements are commonly an order of magnitude (10 times) or considerably more. Tetrode transistors (whether bipolar junction or field-effect) do not exhibit improvements of such a great degree.

The FinFET is a double-gate silicon-on-insulator device, one of a number of geometries being introduced to mitigate the effects of short channels and reduce drain-induced barrier lowering. The *fin* refers to the narrow channel between source and drain. A thin insulating oxide layer on either side of the fin separates it from the gate. SOI FinFETs with a thick oxide on top of the fin are called *double-gate* and those with a thin oxide on top as well as on the sides are called *triple-gate* FinFETs.

A double-gate MOSFET transistor was first demonstrated in 1984 by Electrotechnical Laboratory researchers Toshihiro Sekigawa and Yutaka Hayashi. A GAAFET (gate-all-around MOSFET), a type of multi-gate non-planar 3D transistor, was first demonstrated in 1988, by a Toshiba research team including Fujio Masuoka, H. Takato and K. Sunouchi. The FinFET (fin field-effect transistor), a type of 3D non-planar double-gate MOSFET, originated from the research of Digh Hisamoto and his team at Hitachi Central Research Laboratory in 1989. The development of nanowire multi-gate MOSFETs have since become fundamental to nanoelectronics.

Radiation-hardened-by-design (RHBD)

Semiconductor sub-micrometer and nanometer electronic circuits are the primary concern for operating within the normal tolerance in harsh radiation environments like outer space. One of the design approaches for making a radiation-hardened-by-design (RHBD) device is enclosed-layout-transistor (ELT). Normally, the gate of the MOSFET surrounds the drain, which is placed in the center of the ELT. The source of the MOSFET surrounds the gate. Another RHBD MOSFET is called H-Gate. Both of these transistors have very low leakage current with respect to radiation. However, they are large in size and take more space on silicon than a standard MOSFET. In older STI (shallow trench isolation) designs, radiation strikes near the silicon oxide region cause the channel inversion at the corners of the standard MOSFET due to accumulation of radiation induced trapped charges. If the charges are large enough, the accumulated charges affect STI surface edges along the channel near the channel interface (gate) of the standard MOSFET. Thus the device channel inversion occurs along the channel edges and the device creates off-state leakage path, causing device to turn on. So the reliability of circuits degrades severely. The ELT offers many advantages. These advantages include improvement of reliability by reducing unwanted surface inversion at the gate edges that occurs in the standard MOSFET. Since the gate edges are enclosed in ELT, there is no gate oxide edge (STI at gate interface), and thus the transistor off-state leakage is reduced very much. Low-power microelectronic circuits including computers, communication devices and monitoring systems in space shuttle and satellites are very different from what is used on earth. They are radiation (high-speed atomic particles like proton and neutron, solar flare magnetic energy dissipation in Earth's space, energetic cosmic rays like X-ray, gamma ray etc.) tolerant circuits. These special electronics are designed by applying different techniques using RHBD MOSFETs to ensure safe space journeys and safe space-walks of astronauts.

Applications

Discrete MOSFETs

The MOSFET is by far the most widely used transistor in both digital circuits and analog circuits, and it is the backbone of modern electronics. A common use in analog circuits is the construction of differential amplifiers, used as input stages in op-amps, video amplifiers, high-speed comparators, and many other analog circuits.

Discrete MOSFET devices are widely used in applications such as switch mode power supplies, variable-frequency drives and other power electronics applications where each device may be switching thousands of watts. Radio-frequency amplifiers up to the UHF spectrum use MOSFET transistors as analog signal and power amplifiers. Radio systems also use MOSFETs as oscillators, or mixers to convert frequencies. MOSFET devices are also applied in audio-frequency power amplifiers for public address systems, sound reinforcement and home and automobile sound systems.

MOSFET devices were adopted by NASA for space research in 1964, for its Interplanetary Monitoring Platform (IMP) program. The use of MOSFETs was a major step forward in spacecraft electronics design. Data gathered by IMP spacecraft and satellites were used to support the Apollo program, enabling the first manned Moon landing with the Apollo 11 mission in 1969.

Power Electronics

The power MOSFET is the most widely used power device in the world. Advantages over bipolar junction transistors in power electronics include MOSFETs not requiring a continuous flow of drive current to remain in the ON state, offering higher switching speeds, lower switching power losses, lower on-resistances, and reduced susceptibility to thermal runaway. The power MOSFET had an impact on power supplies, enabling higher operating frequencies, size and weight reduction, and increased volume production.

MOSFETs are commonly used in switched-mode power supplies (SMPS) and DC-to-DC converters. Other power electronic applications include portable information appliances, power integrated circuits, cell phones, notebook computers, and the communications infrastructure that enables the Internet. Switching power supplies are the most common applications for power MOSFETs. They are also widely used for pulsed DC motor drives and Class D audio amplifiers.

MOSFETs are the most widely used RF power amplifiers. MOSFET-based RF power amplifiers enabled the transition of mobile networks from analog to digital in the 1990s, leading to the wide proliferation of wireless mobile networks, which revolutionised telecommunication systems. The LDMOS in particular is the most widely used power amplifier in mobile networks, such as 2G, 3G, and 4G.

Over 50 billion discrete power MOSFETs are shipped annually as of 2018. They are widely used for automotive, industrial and communications systems in particular. Power MOSFETs are commonly used in automotive electronics, particularly as switching devices in electronic control units.

2D Electron Gas

The MOSFET enables physicists to study electron behavior in a two-dimensional gas. In a MOSFET, conduction electrons travel in a thin surface layer, and a "gate" voltage controls the number of charge carriers in this layer. This allows researchers to explore quantum effects by operating high-purity MOSFETs at liquid helium temperatures.

In 1978, the Gakushuin University researchers Jun-ichi Wakabayashi and Shinji Kawaji observed the Hall effect in experiments carried out on the inversion layer of MOSFETs. In 1980, Klaus von

Klitzing, working at the high magnetic field laboratory in Grenoble with silicon-based MOSFET samples developed by Michael Pepper and Gerhard Dorda, made the unexpected discovery of the quantum Hall effect. For this finding, von Klitzing was awarded the 1985 Nobel Prize in Physics.

MOS Integrated Circuits

Following the development of clean rooms to reduce contamination to levels never before thought necessary, and of photolithography and the planar process to allow circuits to be made in very few steps, the Si–SiO_2 system possessed the technical attractions of low cost of production (on a per circuit basis) and ease of integration. Largely because of these two factors, the MOSFET has become the most widely used type of transistor in integrated circuits (ICs), and it is the most critical device component in modern ICs.

In 1961, the MOSFET's co-inventor Dawon Kahng proposed the concept of an MOS integrated circuit, noting that the MOS transistor's ease of fabrication made it useful for integrated circuits. The earliest experimental MOS IC to be demonstrated was a 16-transistor chip built by Fred Heiman and Steven Hofstein at RCA in 1962. General Microelectronics later introduced the first commercial MOS integrated circuit in 1964; it consisted of 120 p-channel transistors, and was a 20-bit shift register, developed by Robert Norman. In 1967, Bell Labs researchers Robert Kerwin, Donald Klein and John Sarace developed the self-aligned gate (silicon-gate) MOS transistor, which Fairchild Semiconductor researchers Federico Faggin and Tom Klein used to develop the first silicon-gate MOS IC.

Large-scale Integration (LSI)

With its high scalability, and much lower power consumption and higher density than bipolar junction transistors, the MOSFET made it possible to build high-density ICs. MOS technology enabled the integration of more than 10,000 transistors in a single LSI (large-scale integration) chip by the early 1970s, and later very large-scale integration (VLSI).

One of the earliest influential consumer electronic products enabled by MOS LSI circuits was the electronic pocket calculator. In 1965, the Victor 3900 desktop calculator was the first MOS LSI calculator, with 29 MOS LSI chips. In 1967 the Texas Instruments Cal-Tech was the first prototype electronic handheld calculator, with three MOS LSI chips, and it was later released as the Canon Pocketronic in 1970. The Sharp QT-8D desktop calculator was the first mass-produced LSI MOS calculator in 1969, and the Sharp EL-8 which used four MOS LSI chips was the first commercial electronic handheld calculator in 1970. The first true electronic pocket calculator was the Busicom LE-120A HANDY LE, which used a single MOS LSI calculator-on-a-chip from Mostek, and was released in 1971.

Microprocessors

The MOSFET is the basis of every microprocessor. The earliest microprocessors were all MOS microprocessors, built with MOS LSI circuits. The first multi-chip microprocessors, the Four-Phase Systems AL1 in 1969 and the Garrett AiResearch MP944 in 1970, were developed with multiple MOS LSI chips. The first commercial single-chip microprocessor, the Intel 4004, was developed by Federico Faggin, using his silicon-gate MOS IC technology, with Intel engineers Marcian Hoff and

Stan Mazor, and Busicom engineer Masatoshi Shima. With the arrival of CMOS microprocessors in 1975, the term "MOS microprocessors" began to refer to chips fabricated entirely from PMOS logic or fabricated entirely from NMOS logic, contrasted with "CMOS microprocessors" and "bipolar bit-slice processors".

CMOS Circuits

Digital

The growth of digital technologies like the microprocessor has provided the motivation to advance MOSFET technology faster than any other type of silicon-based transistor. A big advantage of MOSFETs for digital switching is that the oxide layer between the gate and the channel prevents DC current from flowing through the gate, further reducing power consumption and giving a very large input impedance. The insulating oxide between the gate and channel effectively isolates a MOSFET in one logic stage from earlier and later stages, which allows a single MOSFET output to drive a considerable number of MOSFET inputs. Bipolar transistor-based logic (such as TTL) does not have such a high fanout capacity. This isolation also makes it easier for the designers to ignore to some extent loading effects between logic stages independently. That extent is defined by the operating frequency: as frequencies increase, the input impedance of the MOSFETs decreases.

Analog

The MOSFET's advantages in digital circuits do not translate into supremacy in all analog circuits. The two types of circuit draw upon different features of transistor behavior. Digital circuits switch, spending most of their time either fully on or fully off. The transition from one to the other is only of concern with regards to speed and charge required. Analog circuits depend on operation in the transition region where small changes to V_{gs} can modulate the output (drain) current. The JFET and bipolar junction transistor (BJT) are preferred for accurate matching (of adjacent devices in integrated circuits), higher transconductance and certain temperature characteristics which simplify keeping performance predictable as circuit temperature varies.

Nevertheless, MOSFETs are widely used in many types of analog circuits because of their own advantages (zero gate current, high and adjustable output impedance and improved robustness vs. BJTs which can be permanently degraded by even lightly breaking down the emitter-base). The characteristics and performance of many analog circuits can be scaled up or down by changing the sizes (length and width) of the MOSFETs used. By comparison, in bipolar transistors the size of the device does not significantly affect its performance. MOSFETs' ideal characteristics regarding gate current (zero) and drain-source offset voltage (zero) also make them nearly ideal switch elements, and also make switched capacitor analog circuits practical. In their linear region, MOSFETs can be used as precision resistors, which can have a much higher controlled resistance than BJTs. In high power circuits, MOSFETs sometimes have the advantage of not suffering from thermal runaway as BJTs do. Also, MOSFETs can be configured to perform as capacitors and gyrator circuits which allow op-amps made from them to appear as inductors, thereby allowing all of the normal analog devices on a chip (except for diodes, which can be made smaller than a MOSFET anyway) to be built entirely out of MOSFETs. This means that complete analog circuits can be made on a silicon chip in a much smaller space and with simpler fabrication techniques. MOSFETS are ideally suited to switch inductive loads because of tolerance to inductive kickback.

Some ICs combine analog and digital MOSFET circuitry on a single mixed-signal integrated circuit, making the needed board space even smaller. This creates a need to isolate the analog circuits from the digital circuits on a chip level, leading to the use of isolation rings and silicon on insulator (SOI). Since MOSFETs require more space to handle a given amount of power than a BJT, fabrication processes can incorporate BJTs and MOSFETs into a single device. Mixed-transistor devices are called bi-FETs (bipolar FETs) if they contain just one BJT-FET and BiCMOS (bipolar-CMOS) if they contain complementary BJT-FETs. Such devices have the advantages of both insulated gates and higher current density.

Analog Switches

MOSFET analog switches use the MOSFET to pass analog signals when on, and as a high impedance when off. Signals flow in both directions across a MOSFET switch. In this application, the drain and source of a MOSFET exchange places depending on the relative voltages of the source/drain electrodes. The source is the more negative side for an N-MOS or the more positive side for a P-MOS. All of these switches are limited on what signals they can pass or stop by their gate-source, gate-drain and source–drain voltages; exceeding the voltage, current, or power limits will potentially damage the switch.

Single-type

This analog switch uses a four-terminal simple MOSFET of either P or N type.

In the case of an n-type switch, the body is connected to the most negative supply (usually GND) and the gate is used as the switch control. Whenever the gate voltage exceeds the source voltage by at least a threshold voltage, the MOSFET conducts. The higher the voltage, the more the MOSFET can conduct. An N-MOS switch passes all voltages less than $V_{gate} - V_{tn}$. When the switch is conducting, it typically operates in the linear (or ohmic) mode of operation, since the source and drain voltages will typically be nearly equal.

In the case of a P-MOS, the body is connected to the most positive voltage, and the gate is brought to a lower potential to turn the switch on. The P-MOS switch passes all voltages higher than $V_{gate} - V_{tp}$ (threshold voltage V_{tp} is negative in the case of enhancement-mode P-MOS).

Dual-type (CMOS)

This "complementary" or CMOS type of switch uses one P-MOS and one N-MOS FET to counteract the limitations of the single-type switch. The FETs have their drains and sources connected in parallel, the body of the P-MOS is connected to the high potential (V_{DD}) and the body of the N-MOS is connected to the low potential (*gnd*). To turn the switch on, the gate of the P-MOS is driven to the low potential and the gate of the N-MOS is driven to the high potential. For voltages between $V_{DD} - V_{tn}$ and $gnd - V_{tp}$, both FETs conduct the signal; for voltages less than $gnd - V_{tp}$, the N-MOS conducts alone; and for voltages greater than $V_{DD} - V_{tn}$, the P-MOS conducts alone.

The voltage limits for this switch are the gate-source, gate-drain and source-drain voltage limits for both FETs. Also, the P-MOS is typically two to three times wider than the N-MOS, so the switch will be balanced for speed in the two directions.

Tri-state circuitry sometimes incorporates a CMOS MOSFET switch on its output to provide for a low-ohmic, full-range output when on, and a high-ohmic, mid-level signal when off.

MOS Memory Cells

The first modern memory cells were introduced in 1965, when John Schmidt designed the first MOS semiconductor memory, a 64-bit MOS SRAM (static random-access memory). In 1967, Robert H. Dennard of IBM filed a patent for a single-transistor DRAM (dynamic random-access memory) memory cell, using a MOSFET. MOS memory enabled higher performance, was cheaper, and consumed less power, than magnetic-core memory, leading to MOS memory overtaking magnetic core memory as the dominant computer memory technology by the early 1970s.

In 1967, Dawon Kahng and Simon Sze proposed that floating-gate memory cells, consisting of floating-gate MOSFETs (FGMOS), could be used to produce reprogrammable ROM (read-only memory). Floating-gate memory cells later became the basis for non-volatile memory (NVM) technologies including EPROM (erasable programable ROM), EEPROM (electrically erasable programmable ROM) and flash memory.

Construction

Gate Material

The primary criterion for the gate material is that it is a good conductor. Highly doped polycrystalline silicon is an acceptable but certainly not ideal conductor, and also suffers from some more technical deficiencies in its role as the standard gate material. Nevertheless, there are several reasons favoring use of polysilicon:

1. The threshold voltage (and consequently the drain to source on-current) is modified by the work function difference between the gate material and channel material. Because polysilicon is a semiconductor, its work function can be modulated by adjusting the type and level of doping. Furthermore, because polysilicon has the same bandgap as the underlying silicon channel, it is quite straightforward to tune the work function to achieve low threshold voltages for both NMOS and PMOS devices. By contrast, the work functions of metals are not easily modulated, so tuning the work function to obtain low threshold voltages (LVT) becomes a significant challenge. Additionally, obtaining low-threshold devices on both PMOS and NMOS devices sometimes requires the use of different metals for each device type. While bimetallic integrated circuits (i.e., one type of metal for gate electrodes of NFETS and a second type of metal for gate electrodes of PFETS) are not common, they are known in patent literature and provide some benefit in terms of tuning electrical circuits' overall electrical performance.

2. The silicon-SiO_2 interface has been well studied and is known to have relatively few defects. By contrast many metal-insulator interfaces contain significant levels of defects which can lead to Fermi level pinning, charging, or other phenomena that ultimately degrade device performance.

3. In the MOSFET IC fabrication process, it is preferable to deposit the gate material prior to certain high-temperature steps in order to make better-performing transistors. Such high

temperature steps would melt some metals, limiting the types of metal that can be used in a metal-gate-based process.

While polysilicon gates have been the de facto standard for the last twenty years, they do have some disadvantages which have led to their likely future replacement by metal gates. These disadvantages include:

1. Polysilicon is not a great conductor (approximately 1000 times more resistive than metals) which reduces the signal propagation speed through the material. The resistivity can be lowered by increasing the level of doping, but even highly doped polysilicon is not as conductive as most metals. To improve conductivity further, sometimes a high-temperature metal such as tungsten, titanium, cobalt, and more recently nickel is alloyed with the top layers of the polysilicon. Such a blended material is called silicide. The silicide-polysilicon combination has better electrical properties than polysilicon alone and still does not melt in subsequent processing. Also the threshold voltage is not significantly higher than with polysilicon alone, because the silicide material is not near the channel. The process in which silicide is formed on both the gate electrode and the source and drain regions is sometimes called salicide, self-aligned silicide.

2. When the transistors are extremely scaled down, it is necessary to make the gate dielectric layer very thin, around 1 nm in state-of-the-art technologies. A phenomenon observed here is the so-called poly depletion, where a depletion layer is formed in the gate polysilicon layer next to the gate dielectric when the transistor is in the inversion. To avoid this problem, a metal gate is desired. A variety of metal gates such as tantalum, tungsten, tantalum nitride, and titanium nitride are used, usually in conjunction with high-κ dielectrics. An alternative is to use fully silicided polysilicon gates, a process known as FUSI.

Present high performance CPUs use metal gate technology, together with high-κ dielectrics, a combination known as *high-κ, metal gate* (HKMG). The disadvantages of metal gates are overcome by a few techniques:

1. The threshold voltage is tuned by including a thin "work function metal" layer between the high-κ dielectric and the main metal. This layer is thin enough that the total work function of the gate is influenced by both the main metal and thin metal work functions (either due to alloying during annealing, or simply due to the incomplete screening by the thin metal). The threshold voltage thus can be tuned by the thickness of the thin metal layer.

2. High-κ dielectrics are now well studied, and their defects are understood.

3. HKMG processes exist that do not require the metals to experience high temperature anneals; other processes select metals that can survive the annealing step.

Insulator

As devices are made smaller, insulating layers are made thinner, often through steps of thermal oxidation or localised oxidation of silicon (LOCOS). For nano-scaled devices, at some point tunneling of carriers through the insulator from the channel to the gate electrode takes place. To reduce the resulting leakage current, the insulator can be made thinner by choosing a material with a higher dielectric constant. To see how thickness and dielectric constant are related, note that Gauss's law connects field to charge as:

$$Q = \kappa\epsilon_0 E,$$

with Q = charge density, κ = dielectric constant, ε_o = permittivity of empty space and E = electric field. From this law it appears the same charge can be maintained in the channel at a lower field provided κ is increased. The voltage on the gate is given by:

$$V_G = V_{ch} + Et_{ins} = V_{ch} + \frac{Qt_{ins}}{\kappa\epsilon_0},$$

with V_G = gate voltage, V_{ch} = voltage at channel side of insulator, and t_{ins} = insulator thickness. This equation shows the gate voltage will not increase when the insulator thickness increases, provided κ increases to keep t_{ins} / κ = constant.

The insulator in a MOSFET is a dielectric which can in any event be silicon oxide, formed by LOCOS but many other dielectric materials are employed. The generic term for the dielectric is gate dielectric since the dielectric lies directly below the gate electrode and above the channel of the MOSFET.

Junction Design

The source-to-body and drain-to-body junctions are the object of much attention because of three major factors: their design affects the current-voltage (*I-V*) characteristics of the device, lowering output resistance, and also the speed of the device through the loading effect of the junction capacitances, and finally, the component of stand-by power dissipation due to junction leakage.

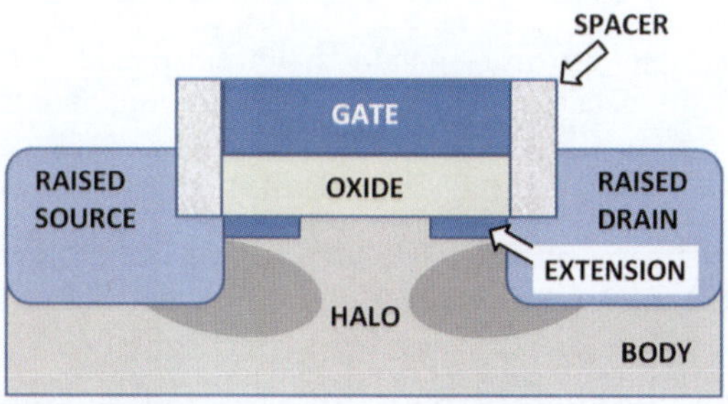

MOSFET showing shallow junction extensions, raised source and drain and halo implant.
Raised source and drain separated from gate by oxide spacers.

The drain induced barrier lowering of the threshold voltage and channel length modulation effects upon *I-V* curves are reduced by using shallow junction extensions. In addition, *halo* doping can be used, that is, the addition of very thin heavily doped regions of the same doping type as the body tight against the junction walls to limit the extent of depletion regions.

The capacitive effects are limited by using raised source and drain geometries that make most of the contact area border thick dielectric instead of silicon. These various features of junction design are shown in the figure.

Scaling

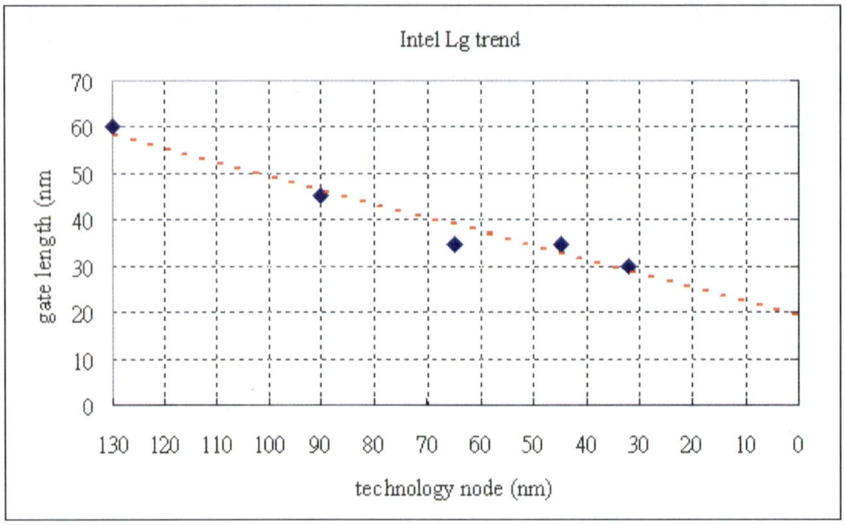

Trend of Intel CPU transistor gate length.

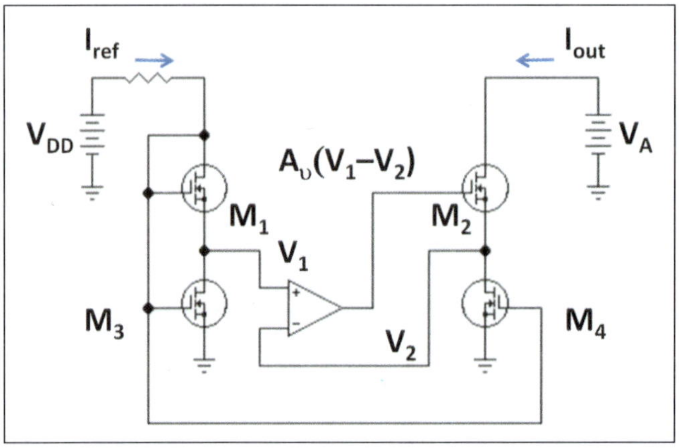

MOSFET version of gain-boosted current mirror; M_1 and M_2 are in active mode, while M_3 and M_4 are in Ohmic mode, and act like resistors. The operational amplifier provides feedback that maintains a high output resistance.

Over the past decades, the MOSFET (as used for digital logic) has continually been scaled down in size; typical MOSFET channel lengths were once several micrometres, but modern integrated circuits are incorporating MOSFETs with channel lengths of tens of nanometers. Robert Dennard's work on scaling theory was pivotal in recognising that this ongoing reduction was possible. The semiconductor industry maintains a "roadmap", the ITRS, which sets the pace for MOSFET development. Historically, the difficulties with decreasing the size of the MOSFET have been associated with the semiconductor device fabrication process, the need to use very low voltages, and with poorer electrical performance necessitating circuit redesign and innovation (small MOSFETs exhibit higher leakage currents and lower output resistance). As of 2019, the smallest MOSFETs in production are 5 nm FinFET semiconductor nodes, manufactured by Samsung Electronics and TSMC.

Smaller MOSFETs are desirable for several reasons. The main reason to make transistors smaller is to pack more and more devices in a given chip area. This results in a chip with the same functionality in a smaller area, or chips with more functionality in the same area. Since fabrication costs for a semiconductor wafer are relatively fixed, the cost per integrated circuits is mainly related to the number of chips that can be produced per wafer. Hence, smaller ICs allow more chips per wafer, reducing the price per chip. In fact, over the past 30 years the number of transistors per chip has been doubled every 2–3 years once a new technology node is introduced. For example, the number of MOSFETs in a microprocessor fabricated in a 45 nm technology can well be twice as many as in a 65 nm chip. This doubling of transistor density was first observed by Gordon Moore in 1965 and is commonly referred to as Moore's law. It is also expected that smaller transistors switch faster. For example, one approach to size reduction is a scaling of the MOSFET that requires all device dimensions to reduce proportionally. The main device dimensions are the channel length, channel width, and oxide thickness. When they are scaled down by equal factors, the transistor channel resistance does not change, while gate capacitance is cut by that factor. Hence, the RC delay of the transistor scales with a similar factor. While this has been traditionally the case for the older technologies, for the state-of-the-art MOSFETs reduction of the transistor dimensions does not necessarily translate to higher chip speed because the delay due to interconnections is more significant.

Producing MOSFETs with channel lengths much smaller than a micrometre is a challenge, and the difficulties of semiconductor device fabrication are always a limiting factor in advancing integrated circuit technology. Though processes such as ALD have improved fabrication for small components, the small size of the MOSFET (less than a few tens of nanometers) has created operational problems.

Higher Subthreshold Conduction

As MOSFET geometries shrink, the voltage that can be applied to the gate must be reduced to maintain reliability. To maintain performance, the threshold voltage of the MOSFET has to be reduced as well. As threshold voltage is reduced, the transistor cannot be switched from complete turn-off to complete turn-on with the limited voltage swing available; the circuit design is a compromise between strong current in the *on* case and low current in the *off* case, and the application determines whether to favor one over the other. Subthreshold leakage (including subthreshold conduction, gate-oxide leakage and reverse-biased junction leakage), which was ignored in the past, now can consume upwards of half of the total power consumption of modern high-performance VLSI chips.

Increased Gate-oxide Leakage

The gate oxide, which serves as insulator between the gate and channel, should be made as thin as possible to increase the channel conductivity and performance when the transistor is on and to reduce subthreshold leakage when the transistor is off. However, with current gate oxides with a thickness of around 1.2 nm (which in silicon is ~5 atoms thick) the quantum mechanical phenomenon of electron tunneling occurs between the gate and channel, leading to increased power consumption. Silicon dioxide has traditionally been used as the gate insulator. Silicon dioxide however has a modest dielectric constant. Increasing the dielectric constant of the gate dielectric allows a thicker layer while maintaining a high capacitance (capacitance is proportional to dielectric constant and inversely proportional to dielectric thickness). All else equal, a higher dielectric thickness reduces the quantum tunneling current through the dielectric between the gate and the channel. Insulators that have a larger dielectric constant than silicon dioxide (referred to as high-κ dielectrics), such as group IVb metal silicates e.g. hafnium and zirconium silicates and oxides are being used to reduce the gate leakage from the 45 nanometer technology node onwards. On the other hand, the barrier height of the new gate insulator is an important consideration; the difference in conduction band energy between the semiconductor and the dielectric (and the corresponding difference in valence band energy) also affects leakage current level. For the traditional gate oxide, silicon dioxide, the former barrier is approximately 8 eV. For many alternative dielectrics the value is significantly lower, tending to increase the tunneling current, somewhat negating the advantage of higher dielectric constant. The maximum gate-source voltage is determined by the strength of the electric field able to be sustained by the gate dielectric before significant leakage occurs. As the insulating dielectric is made thinner, the electric field strength within it goes up for a fixed voltage. This necessitates using lower voltages with the thinner dielectric.

Increased Junction Leakage

To make devices smaller, junction design has become more complex, leading to higher doping levels, shallower junctions, "halo" doping and so forth, all to decrease drain-induced barrier lowering. To keep these complex junctions in place, the annealing steps formerly used to remove damage and electrically active defects must be curtailed increasing junction leakage. Heavier doping is also associated with thinner depletion layers and more recombination centers that result in increased leakage current, even without lattice damage.

Drain-induced barrier lowering (DIBL) and V_T roll off because of the short-channel effect, channel formation is not entirely done by the gate, but now the drain and source also affect the channel formation. As the channel length decreases, the depletion regions of the source and drain come closer together and make the threshold voltage (V_T) a function of the length of the channel. This is called V_T roll-off. V_T also becomes function of drain to source voltage V_{DS}. As we increase the V_{DS}, the depletion regions increase in size, and a considerable amount of charge is depleted by the V_{DS}. The gate voltage required to form the channel is then lowered, and thus, the V_T decreases with an increase in V_{DS}. This effect is called drain induced barrier lowering (DIBL).

Lower Output Resistance

For analog operation, good gain requires a high MOSFET output impedance, which is to say, the MOSFET current should vary only slightly with the applied drain-to-source voltage. As devices are made smaller, the influence of the drain competes more successfully with that of the gate due to the growing proximity of these two electrodes, increasing the sensitivity of the MOSFET current to the drain voltage. To counteract the resulting decrease in output resistance, circuits are made more complex, either by requiring more devices, for example the cascode and cascade amplifiers, or by feedback circuitry using operational amplifiers.

Lower Transconductance

The transconductance of the MOSFET decides its gain and is proportional to hole or electron mobility (depending on device type), at least for low drain voltages. As MOSFET size is reduced, the fields in the channel increase and the dopant impurity levels increase. Both changes reduce the carrier mobility, and hence the transconductance. As channel lengths are reduced without proportional reduction in drain voltage, raising the electric field in the channel, the result is velocity saturation of the carriers, limiting the current and the transconductance.

Interconnect Capacitance

Traditionally, switching time was roughly proportional to the gate capacitance of gates. However, with transistors becoming smaller and more transistors being placed on the chip, interconnect capacitance (the capacitance of the metal-layer connections between different parts of the chip) is becoming a large percentage of capacitance. Signals have to travel through the interconnect, which leads to increased delay and lower performance.

Heat Production

The ever-increasing density of MOSFETs on an integrated circuit creates problems of substantial localized heat generation that can impair circuit operation. Circuits operate more slowly at high temperatures, and have reduced reliability and shorter lifetimes. Heat sinks and other cooling devices and methods are now required for many integrated circuits including microprocessors. Power MOSFETs are at risk of thermal runaway. As their on-state resistance rises with temperature, if the load is approximately a constant-current load then the power loss rises correspondingly, generating further heat. When the heatsink is not able to keep the temperature low enough, the junction temperature may rise quickly and uncontrollably, resulting in destruction of the device.

Process Variations

With MOSFETs becoming smaller, the number of atoms in the silicon that produce many of the transistor's properties is becoming fewer, with the result that control of dopant numbers and placement is more erratic. During chip manufacturing, random process variations affect all transistor dimensions: length, width, junction depths, oxide thickness *etc.*, and become a greater percentage of overall transistor size as the transistor shrinks. The transistor characteristics become less certain, more statistical. The random nature of manufacture means we do not know which particular example MOSFETs actually will end up in a particular instance of the circuit. This uncertainty forces a less optimal design because the design must work for a great variety of possible component MOSFETs.

Modeling Challenges

Modern ICs are computer-simulated with the goal of obtaining working circuits from the very first manufactured lot. As devices are miniaturized, the complexity of the processing makes it difficult to predict exactly what the final devices look like, and modeling of physical processes becomes more challenging as well. In addition, microscopic variations in structure due simply to the probabilistic nature of atomic processes require statistical (not just deterministic) predictions. These factors combine to make adequate simulation and "right the first time" manufacture difficult.

References

- Sungook, Hong (2001). Wireless: From Marconi's Black-Box to the Audion. MIT Press. P. 165. ISBN 978-0262082983

- Introduction-analog-electronics: engineersgarage.com, Retrieved 19 April, 2019
- Agarwal, Anant; Lang, Jeffrey (2005). Foundations of Analog and Digital Electronic Circuits. Morgan Kaufmann. P. 331. ISBN 978-0080506814
- Voltage+amplifier: thefreedictionary.com, Retrieved 17 May, 2019
- Peter Ashburn (2003). Sige Heterojunction Bipolar Transistors. New York: Wiley. Chapter 10. ISBN 978-0-470-84838-8
- Power-amplifier: electronicshub.org, Retrieved 14 July, 2019
- A.S. Sedra and K.C. Smith (2004). Microelectronic Circuits (5th ed.). New York: Oxford. P. 509. ISBN 978-0-19-514251-8
- "IC Temperature Sensors Find the Hot Spots - Application Note - Maxim". Maxim-ic.com. February 21, 2002. Retrieved August 10,2016
- Current-amplifiers-and-buffers: electronicshub.org, Retrieved 31 March, 2019
- Moskowitz, Sanford L. (2016). Advanced Materials Innovation: Managing Global Technology in the 21st century. John Wiley & Sons. P. 168. ISBN 9780470508923

3

Digital Electronics

The electronics that operate on digital signals are known as digital electronics. They are less susceptible to noise and signal attenuation. A few devices which use digital electronics are multiplexers and demultiplexers. All the diverse principles of digital electronics have been carefully analyzed in this chapter.

Digital electronics are those electronics systems that use a digital signal instead of an analog signal. Digital electronics are the most common representation of Boolean algebra and are the basis of all digital circuits for computers, mobile phones, and numerous other consumer products.

A digital circuit that acts as a binary clock, hand-wired on a series of breadboards.

The most common fundamental unit of digital electronics is the logic gate. By combining numerous logic gates (from tens to hundreds of thousands) more complex systems can be created. The complex system of digital electronics is collectively referred to as a digital circuit.

To most electronic engineers, the terms "digital circuit", "digital system" and "logic" are interchangeable in the context of digital circuits.

Advantages

The usual advantages of digital circuits when compared to analog circuits are:

Digital systems interface well with computers and are easy to control with software. It is often possible to add new features to a digital system without changing hardware, and to do this remotely, just by uploading new software. Design errors or bugs can be worked-around with a software upgrade, after the product is in customer hands.

Information storage can be much easier in digital systems than in analog ones. In particular, the great noise-immunity of digital systems makes it possible to store data and retrieve it later without degradation. In an analog system, aging and wear and tear will degrade the information in storage, but in a digital system, as long as the wear and tear is below a certain level, the information can be recovered perfectly.

Robustness

One of the primary advantages of digital electronics stems from it simply being digital: robustness.

The robustness stems from the fact that if the noise is less than the noise margin then the system performs as if there were no noise at all. This is a necessary and sufficient property for a circuit to be considered digital. However, if the noise exceeds this level, the circuit can give unexpected and undesired results — perhaps catastrophic incorrect results.

Digital signals can thus be regenerated to achieve lossless data transmission, within certain limits. Analog signal transmission and processing, by contrast, always introduces noise.

Theoretically, there is no data-loss when copying digital data. This is a great advantage over analog systems, which faithfully reproduce every bit of noise that makes its way into the signal.

Disadvantages

Digital circuits use more energy than analog circuits to accomplish the same calculations and signal processing tasks, thus producing more heat as well. In portable or battery-powered systems this can be a major limiting factor, but in a situation where power is plentiful, a digital system is often preferred because of all the advantages listed above, especially that of (re-)programmability and ease of upgrading without requiring hardware changes.

A particular example is the cellular telephone, which being a battery-powered portable device, uses a low-power analog front-end to acquire and tune in the radio signal from the base station. The base station, being in a fixed location with access to the power grid, can afford to use power-hungry software-defined (digital) radio techniques that digitize the signal essentially at the antenna (after wideband filtering and downconversion to intermediate frequency) and performs all channelization and demodulation via software-driven calculations. Such base stations can be reprogrammed, potentially via remote control, to process the signals used in future cellular standards as those standards become available.

Digital circuits are sometimes more expensive, especially in small quantities.

The world in which we live is analog, and signals from this world such as light, temperature, sound, electrical conductivity, electric and magnetic fields, and phenomena such as the flow of time, are for most practical purposes continuous and thus analog quantities rather than discrete digital ones. For a digital system to do useful things in the real world, translation from the continuous realm to the discrete digital realm must occur, resulting in quantization errors. This problem can usually be mitigated by designing the system to store enough digital data to represent the signal to the desired degree of fidelity. The Nyquist-Shannon sampling theorem provides an important guideline as to how much digital data is needed to accurately portray a given analog signal.

Fragility

Digital systems can be fragile, in that if a single piece of digital data is lost or misinterpreted, the meaning of large blocks of related data can completely change. This problem can be mitigated by designing the digital system for robustness. For example, a parity bit or other error-detecting or error-correcting code can be inserted into the signal path so that if less than a certain fraction of the data is corrupted, the system can determine that this has occurred and possibly uncorrupt the data, or ask for the corrupted data to be resent. In a state-machine, the state transition logic can be designed to catch all unused states and trigger a reset sequence or other error recovery routine.

For example, it is standard practice in embedded software design to fill unused program memory with interrupt instructions that point to an error recovery routine, to help guard against a failure that corrupts the microcontroller's instruction pointer which could otherwise cause random code to be executed.

Analog Issues in Digital Circuits

Digital circuits are made from analog components, and care has to be taken in design so that the analog nature of these underlying components don't dominate over the desired digital behavior. In particular, attention must be paid to all noise and timing margins, to parasitic inductances and capacitances, to proper filtering of power and ground connections, to electromagnetic coupling amongst datalines, and many other details. Inattention to these can cause intermittent problems such as "glitches", vanishingly-fast pulses that may trigger some logic but not others, "runt pulses" that do not reach valid switching (threshold) voltages, or unexpected ("undecoded") combinations of logic states.

A corollary of the fact that digital circuits are made from analog components is the fact that digital circuits are slower to perform calculations than analog circuits that occupy a similar amount of physical space and consume the same amount of power. However, the digital circuit will perform the calculation with much better repeatability, due to the high noise immunity of digital circuitry.

Construction

A digital circuit is often constructed from small electronic circuits called logic gates. Each logic gate represents a function of boolean logic. A logic gate is an arrangement of electrically controlled switches. The output is an electrical flow or voltage, that can, in turn, control more logic gates. Logic gates often use the fewest number of transistors in order to reduce their size, power consumption and cost, and increase their reliability. Manufactured as integrated circuits, they are the least expensive implementation when made in large volumes. They are usually designed by engineers using electronic design automation software.

Another form of digital circuit is constructed from lookup tables, (many sold as "programmable logic devices", though other kinds of PLDs exist). Lookup tables can perform all the same functions as machines based on logic gates, but lookup tables can be easily reprogrammed without changing the wiring. This means that a designer can often repair errors without changing the arrangement of wires. Therefore, in small volume products, programmable logic devices are often the preferred solution. They are usually designed by engineers using electronic design automation software.

When the volumes are medium to large, and the logic can be slow, or involves complex algorithms or sequences, often a small microcontroller is programmed to make an embedded system. These are usually programmed by software engineers.

When only one digital circuit is needed, and its design is totally customized, as for a factory production line controller, the conventional solution is a programmable logic controller, or PLC. These are usually programmed by electricians, using ladder logic.

Structure of Digital Systems

Engineers use many methods to minimize logic functions, in order to reduce the complexity, and thus the number of errors and the expense of digital circuits. The most widely used method is the application of a minimization algorithm within a CAD system, although historically, truth tables, Karnaugh Maps, and Boolean algebra have also been used.

Representations are crucial to an engineer's design of digital circuits. Some analysis methods only work with particular representations.

The classical way to represent a digital circuit is with an equivalent set of logic gates. Another way, often with the least electronics, is to construct an equivalent system of electronic switches (usually transistors). One of the easiest ways is to simply have a memory containing a truth table. The inputs are fed into the address of the memory, and the data outputs of the memory become the outputs.

For automated analysis, these representations have digital file formats that can be processed by computer programs. Most digital engineers are very careful to select computer programs ("tools") with compatible file formats.

To choose representations, engineers consider types of digital systems. Most digital systems divide into "combinatorial systems" and "sequential systems". A combinatorial system always presents the same output when given the same inputs. It is basically a representation of a set of logic functions.

A sequential system is a combinatorial system with some of the outputs fed back as inputs. This makes the digital machine perform a "sequence" of operations. The simplest sequential system is probably a flip flop, a mechanism that represents a binary digit or "bit".

Sequential systems are often designed as state machines. In this way, engineers can design a system's gross behavior, and even test it in a simulation, without considering all the details of the logic functions.

Sequential systems divide into two further subcategories. "Synchronous" sequential systems change state all at once, when a "clock" signal changes state. "Asynchronous" sequential systems propagate changes whenever inputs change. Synchronous sequential systems are made of well-characterized asynchronous circuits such as flip-flops, that change only when the clock changes, and which have carefully designed timing margins.

The usual way to implement a synchronous sequential state machine is divide it into a piece of combinatorial logic and a set of flip flops called a "state register." Each time a clock signal ticks, the state register captures the feedback generated from the previous state of the combinatorial logic, and feeds it back as an unchanging input to the combinatorial part of the state machine. The fastest rate of the clock is set by the most time-consuming logic calculation in the combinatorial logic.

The state register is just a representation of a binary number. If the states in the state machine are numbered (easy to arrange), the logic function is just some logic that produces the number of the next state.

In comparison, asynchronous systems are very hard to design because all possible states, in all possible timings must be considered. The usual method is to construct a table of the minimum and maximum time that each such state can exist, and then adjust the circuit to minimize the number of such states, and force the circuit to periodically wait for all of its parts to enter a compatible state. (This is called "self-resynchronization.") Without such careful design, it is easy to accidentally produce asynchronous logic that is "unstable", that is, real electronics will have unpredictable results because of the cumulative delays caused by small variations in the values of the electronic components. Certain circuits (such as the synchronizer flip-flops, switch debouncers, and the like which allow external unsynchronized signals to enter synchronous logic circuits) are inherently asynchronous in their design and must be analyzed as such.

As of now, almost all digital machines are synchronous designs because it is much easier to create and verify a synchronous design. However, asynchronous logic is thought to be superior, if it can be made to work, because its speed is not constrained by an arbitrary clock; instead, it simply runs at the maximum speed permitted by the propagation rates of the logic gates from which it is constructed. Building an asynchronous circuit using faster parts implicitly makes the circuit "go" faster.

More generally, many digital systems are data flow machines. These are usually designed using synchronous register transfer logic, using hardware description languages such as VHDL or Verilog.

In register transfer logic, binary numbers are stored in groups of flip flops called registers. The outputs of each register are a bundle of wires called a "bus" that carries that number to other calculations. A calculation is simply a piece of combinatorial logic. Each calculation also has an output bus, and these may be connected to the inputs of several registers. Sometimes a register will have a multiplexer on its input, so that it can store a number from any one of several buses. Alternatively, the outputs of several items may be connected to a bus through buffers that can turn off the output of all of the devices except one. A sequential state machine controls when each register accepts new data from its input.

In the 1980s, some researchers discovered that almost all synchronous register-transfer machines could be converted to asynchronous designs by using first-in-first-out synchronization logic. In this scheme, the digital machine is characterized as a set of data flows. In each step of the flow, an asynchronous "synchronization circuit" determines when the outputs of that step are valid, and presents a signal that says, "grab the data" to the stages that use that stage's inputs. It turns out that just a few relatively simple synchronization circuits are needed.

The most general-purpose register-transfer logic machine is a computer. This is basically an automatic binary abacus. The control unit of a computer is usually designed as a microprogram run by a microsequencer. A microprogram is much like a player-piano roll. Each table entry or "word" of the microprogram commands the state of every bit that controls the computer. The sequencer then counts, and the count addresses the memory or combinatorial logic machine that contains the microprogram. The bits from the microprogram control the arithmetic logic unit, memory and other parts of the computer, including the microsequencer itself.

In this way, the complex task of designing the controls of a computer is reduced to a simpler task of programming a relatively independent collection of much simpler logic machines.

Computer architecture is a specialized engineering activity that tries to arrange the registers, calculation logic, buses and other parts of the computer in the best way for some purpose. Computer architects have applied large amounts of ingenuity to computer design to reduce the cost and increase the speed and immunity to programming errors of computers. An increasingly common goal is to reduce the power used in a battery-powered computer system, such as a cell-phone. Many computer architects serve an extended apprenticeship as microprogrammers.

"Specialized computers" are usually a conventional computer with a special-purpose microprogram.

Automated Design Tools

To save costly engineering effort, much of the effort of designing large logic machines has been automated. The computer programs are called "electronic design automation tools" or just "EDA."

Simple truth table-style descriptions of logic are often optimized with EDA that automatically produces reduced systems of logic gates or smaller lookup tables that still produce the desired outputs.

Most practical algorithms for optimizing large logic systems use algebraic manipulations or binary decision diagrams, and there are promising experiments with genetic algorithms and annealing optimizations.

To automate costly engineering effort, some EDA can take state tables that describe state machines and automatically produce a truth table for the combinatorial part of a state machine. The state table is a piece of text that lists each state, and the conditions that can exit that state.

It is common for the truth tables of such computer-generated state-machines to be optimized with logic-minimization software. This is a simple example of how complex logic machines are broken into simpler parts. Often, real logic systems are designed as a series of sub-projects, which are combined using a "tool flow". The tool flow is usually a "script", a simplified computer language that can invoke the software design tools in the right order.

Tool flows for large logic systems such as microprocessors can be thousands of commands long, and combine the work of hundreds of engineers.

Writing and debugging tool flows is an established engineering specialty in companies that produce complex logic machines. The tool flow usually terminates in a detailed computer file or set of files that describe how to physically construct the logic machine. Often it consists of instructions to draw the transistors and wires on an integrated circuit or a printed circuit board.

Parts of tool flows are "debugged" by testing the outputs of simulated logic machines against expected inputs. The test tools take computer files with sets of inputs and outputs, and highlight discrepancies between the simulated behavior and the expected behavior.

These test data are usually called "test vectors." Often, the test vectors are preserved and used in the factory to test that newly constructed logic machines work correctly.

Design for Testability

A large logic machine (say, with more than a hundred logical variables) can have an astronomical number of possible states. Obviously, in the factory, testing every state is impractical if testing each state takes a microsecond, and there are more states than the number of microseconds since the universe began. Unfortunately, this ridiculous-sounding case is typical.

Fortunately, large logic machines are almost always designed as assemblies of smaller logic machines. To save time, the smaller sub-machines are isolated by permanently-installed "design for test" circuitry, and are tested independently.

One common test scheme known as "scan design" moves test bits serially (one after another) from external test equipment through one or more serial shift registers known as "scan chains". Serial scans have only one or two wires to carry the data, and minimize the physical size and expense of the infrequently-used test logic.

After all the test data bits are in place, the design is reconfigured to be in "normal mode" and one or more clock pulses are applied, to test for faults (e.g. stuck-at low or stuck-at high) and capture the test result into flip-flops and/or latches in the scan shift register(s). Finally, the result of the test is shifted out to the block boundary and compared against the predicted "good machine" result.

In a board-test environment, serial to parallel testing has been formalized with a standard called "JTAG" (named after the "Joint Testing Acting Group" that proposed it).

Another common testing scheme provides a test mode that forces some part of the logic machine to enter a "test cycle." The test cycle usually exercises large independent parts of the machine.

Trade-offs

Several numbers determine the practicality of a system of digital logic. Engineers explored numerous electronic devices to get an ideal combination of speed, low cost and reliability.

The cost of a logic gate is crucial. In the 1930s, the earliest digital logic systems were constructed from telephone relays because these were inexpensive and relatively reliable. After that, engineers always used the cheapest available electronic switches that could still fulfill the requirements.

The earliest integrated circuits were a happy accident. They were constructed not to save money, but to save weight, and permit the Apollo Guidance Computer to control an inertial guidance system for a spacecraft. The first integrated circuit logic gates cost nearly $50 (in 1960 dollars, when an engineer earned $10,000/year). To everyone's surprise, by the time the circuits were mass-produced, they had become the least-expensive method of constructing digital logic. Improvements in this technology have driven all subsequent improvements in cost.

With the rise of integrated circuits, reducing the absolute number of chips used represented another way to save costs. The goal of a designer is not just to make the simplest circuit, but to keep the component count down. Sometimes this results in slightly more complicated designs with respect to the underlying digital logic but nevertheless reduces the number of components, board size, and even power consumption.

For example, in some logic families, NAND gates are the simplest digital gate to build. All other logical operations can be implemented by NAND gates. If a circuit already required a single NAND gate, and a single chip normally carried four NAND gates, then the remaining gates could be used to implement other logical operations like logical and. This could eliminate the need for a separate chip containing those different types of gates.

The "reliability" of a logic gate describes its mean time between failure (MTBF). Digital machines often have millions of logic gates. Also, most digital machines are "optimized" to reduce their cost. The result is that often, the failure of a single logic gate will cause a digital machine to stop working.

Digital machines first became useful when the MTBF for a switch got above a few hundred hours. Even so, many of these machines had complex, well-rehearsed repair procedures, and would be nonfunctional for hours because a tube burned-out, or a moth got stuck in a relay. Modern transistorized integrated circuit logic gates have MTBFs of nearly a trillion (1x10^12) hours, and need them because they have so many logic gates.

The "fan out" describes how many logic inputs can be controlled by a single logic output. The minimum practical fan out is about five. Modern electronic logic using CMOS transistors for switches have fanouts near fifty, and can sometimes go much higher.

The "switching speed" describes how many times per second an inverter (an electronic representation of a "logical not" function) can change from true to false and back. Faster logic can accomplish more operations in less time. Digital logic first became useful when switching speeds got above fifty hertz, because that was faster than a team of humans operating mechanical calculators. Modern electronic digital logic routinely switches at five gigahertz ($5x10^9$ hertz), and some laboratory systems switch at more than a terahertz ($1x10^{12}$ hertz).

Non-electronic Logic

It is possible to construct non-electronic digital mechanisms. In principle, any technology capable of representing discrete states and representing logic operations could be used to build mechanical logic.

Hydraulic, pneumatic and mechanical versions of logic gates exist and are used in situations where electricity cannot be used. The first two types are considered under the heading of fluidics. One application of fluidic logic is in military hardware that is likely to be exposed to a nuclear electromagnetic pulse (nuclear EMP, or NEMP) that would destroy any electrical circuits.

Mechanical logic is frequently used in inexpensive controllers, such as those in washing machines. Famously, the first computer design, by Charles Babbage, was designed to use mechanical logic. Mechanical logic might also be used in very small computers that could be built by nanotechnology.

Another example is that if two particular enzymes are required to prevent the construction of a particular protein, this is the equivalent of a biological "NAND" gate. The term digital signal is used to refer to more than one concept. It can refer to discrete-time signals that are digitized, or to the waveform signals in a digital system.

Discrete-time Signals

Template: Main Digital signals are digital representations of discrete-time signals, which are often derived from analog signals.

An analog signal is a datum that changes over time—say, the temperature at a given location; the depth of a certain point in a pond; or the amplitude of the voltage at some node in a circuit—that can be represented as a mathematical function, with time as the free variable (abscissa) and the signal itself as the dependent variable (ordinate). A discrete-time signal is a sampled version of an analog signal: the value of the datum is noted at fixed intervals (for example, every microsecond) rather than continuously.

If individual time values of the discrete-time signal, instead of being measured precisely (which would require an infinite number of digits), are approximated to a certain precision—which, therefore, only requires a specific number of digits—then the resultant data stream is termed a digital signal. The process of approximating the precise value within a fixed number of digits, or bits, is called quantization.

In conceptual summary, a digital signal is a quantized discrete-time signal; a discrete-time signal is a sampled analog signal.

In the Digital Revolution, the usage of digital signals has increased significantly. Many modern media devices, especially the ones that connect with computers use digital signals to represent signals that were traditionally represented as continuous-time signals; cell phones, music and video players, personal video recorders, and digital cameras are examples.

In most applications, digital signals are represented as binary numbers, so their precision of quantization is measured in bits. Suppose, for example, that we wish to measure a signal to two significant decimal digits. Since seven bits, or binary digits, can record 128 discrete values (viz., from 0 to 127), those seven bits are more than sufficient to express a range of one hundred values.

Waveforms in Digital Systems

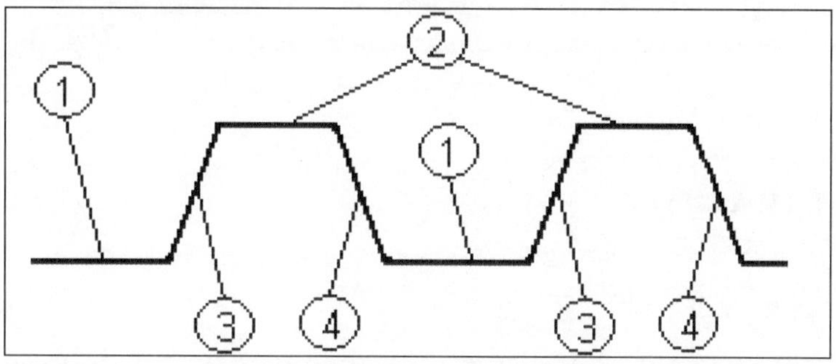

Digital signal: 1) Low level, 2) High level, 3) Rising edge, and 4) Falling edge.

In computer architecture and other digital systems, a waveform that switches between two voltage levels representing the two states of a Boolean value (0 and 1) is referred to as a digital signal, even though it is an analog voltage waveform, since it is interpreted in terms of only two levels.

The clock signal is a special digital signal that is used to synchronize digital circuits. The image shown can be considered the waveform of a clock signal. Logic changes are triggered either by the rising edge or the falling edge.

Logic Voltage Levels

The two states of a wire are usually represented by some measurement of electric current: Voltage is the most common, but current is used in some logic families. A threshold is designed for each logic family. When below that threshold, the wire is "low," when above "high." Digital circuits establish a "no man's area" or "exclusion zone" that is wider than the tolerances of the components. The circuits avoid that area, in order to avoid indeterminate results.

It is usual to allow some tolerance in the voltage levels used; for example, 0 to 2 volts might represent logic 0, and 3 to 5 volts logic 1. A voltage of 2 to 3 volts would be invalid and would occur only in a fault condition or during a logic level transition, as most circuits are not purely resistive, and therefore cannot instantly change voltage levels. However, few logic circuits can detect such a fault, and most will just choose to interpret the signal randomly as either a 0 or a 1.

The levels represent the binary integers or logic levels of 0 and 1. In active-high logic, "low" represents binary 0 and "high" represents binary 1. Active-low logic uses the reverse representation.

Hobbyist frequency counter circuit built almost entirely of TTL logic chips.

Digital Logic Gates

There are three basic types of digital logic gates, the AND Gate, the OR Gate and the NOT Gate

Digital Logic Gates can be made from discrete components such as Resistors, Transistors and Diodes to form RTL (resistor-transistor logic) or DTL (diode-transistor logic) circuits, but today's modern digital 74xxx series integrated circuits are manufactured using TTL (transistor-transistor logic) based on NPN bipolar transistor technology or the much faster and low power CMOS based MOSFET transistor logic used in the 74Cxxx, 74HCxxx, 74ACxxx and the 4000 series logic chips.

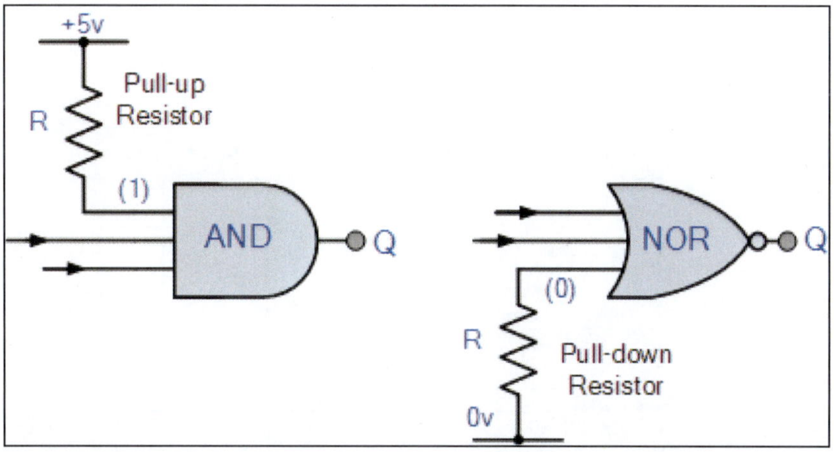

The eight most "standard" individual Digital Logic Gates are summarised below along with their corresponding truth tables.

Standard Logic Gates

The Logic AND Gate

Symbol	Truth Table		
 2-input AND Digital Logic Gate	B	A	Q
	0	0	0
	0	1	0
	1	0	0
	1	1	1
Boolean Expression Q = A.B	Read as A AND B gives Q		

The Logic OR Gate

Symbol	Truth Table		
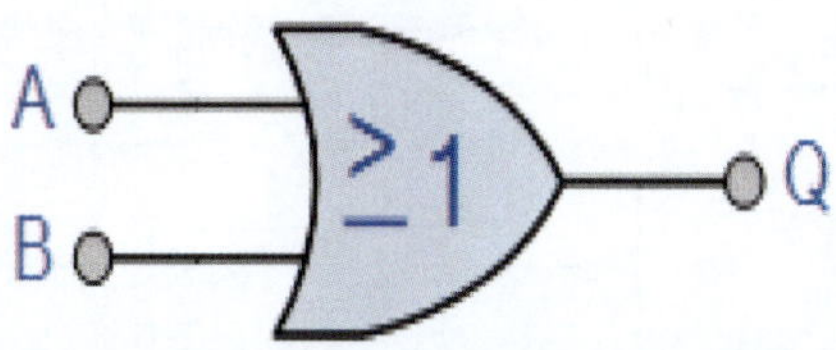	B	A	Q
	0	0	0
	0	1	1
	1	0	1
	1	1	1
Boolean Expression Q = A + B	Read as A OR B gives Q		

Inverting Logic Gates

The Logic NAND Gate

<table>
<tr><th>Symbol</th><th colspan="3">Truth Table</th></tr>
<tr><td rowspan="5">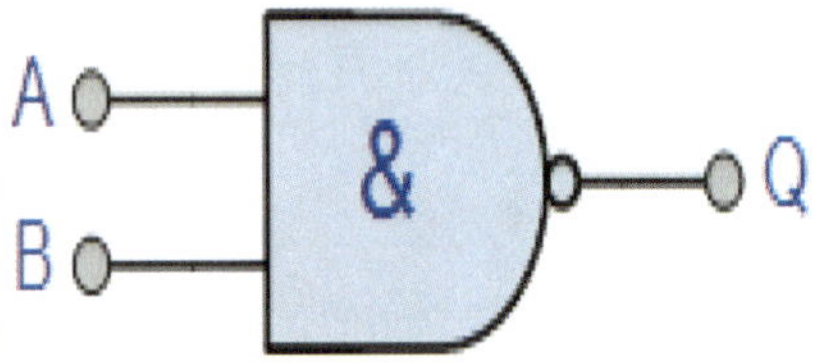
</td><td>B</td><td>A</td><td>Q</td></tr>
<tr><td>0</td><td>0</td><td>1</td></tr>
<tr><td>0</td><td>1</td><td>1</td></tr>
<tr><td>1</td><td>0</td><td>1</td></tr>
<tr><td>1</td><td>1</td><td>0</td></tr>
<tr><td>Boolean Expression Q = A . B</td><td colspan="3">Read as A AND B gives NOT Q</td></tr>
</table>

The Logic NOR Gate

<table>
<tr><th>Symbol</th><th colspan="3">Truth Table</th></tr>
<tr><td rowspan="5">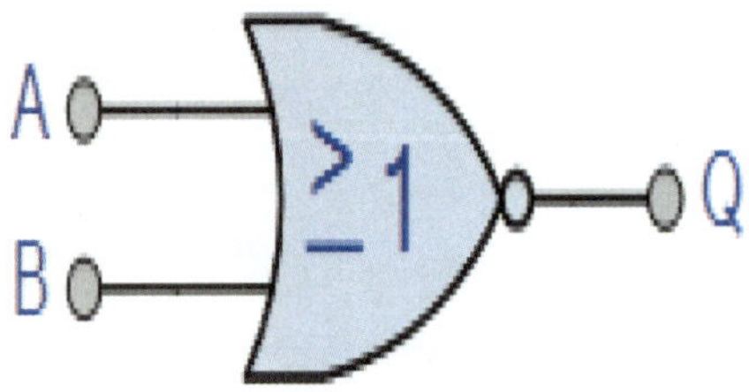
</td><td>B</td><td>A</td><td>Q</td></tr>
<tr><td>0</td><td>0</td><td>1</td></tr>
<tr><td>0</td><td>1</td><td>0</td></tr>
<tr><td>1</td><td>0</td><td>0</td></tr>
<tr><td>1</td><td>1</td><td>0</td></tr>
<tr><td>Boolean Expression Q = A + B</td><td colspan="3">Read as A OR B gives NOT Q</td></tr>
</table>

Exclusive Logic Gates

The Logic Exclusive-OR Gate (Ex-OR)

<table>
<tr><th>Symbol</th><th colspan="3">Truth Table</th></tr>
<tr><td rowspan="5">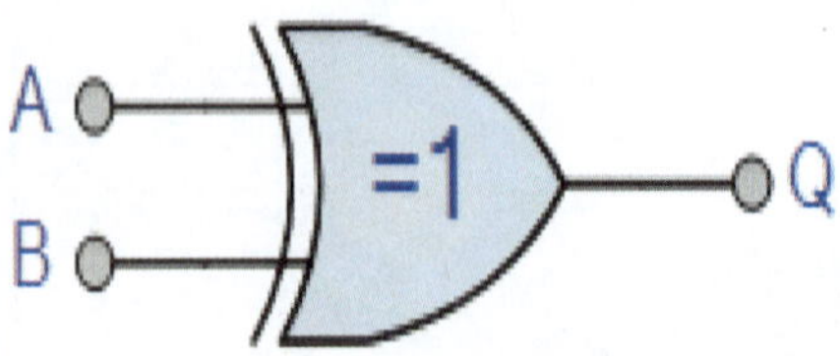
</td><td>B</td><td>A</td><td>Q</td></tr>
<tr><td>0</td><td>0</td><td>0</td></tr>
<tr><td>0</td><td>1</td><td>1</td></tr>
<tr><td>1</td><td>0</td><td>1</td></tr>
<tr><td>1</td><td>1</td><td>0</td></tr>
<tr><td>Boolean Expression Q = A ⊕ B</td><td colspan="3">Read as A OR B but not BOTH gives Q (odd)</td></tr>
</table>

The Logic Exclusive-NOR Gate (Ex-NOR)

<table>
<tr><th>Symbol</th><th colspan="3">Truth Table</th></tr>
<tr><td rowspan="5">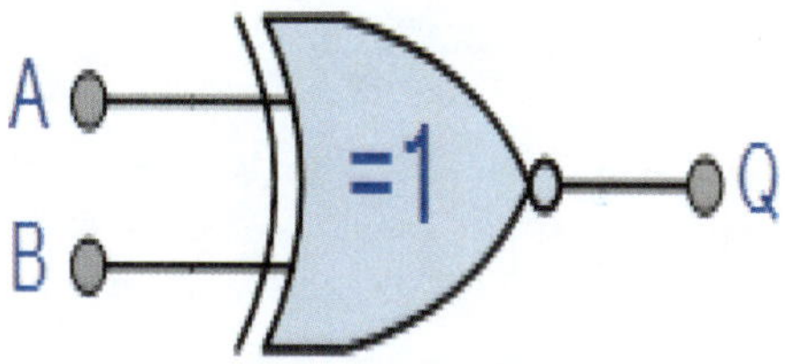
</td><td>B</td><td>A</td><td>Q</td></tr>
<tr><td>0</td><td>0</td><td>1</td></tr>
<tr><td>0</td><td>1</td><td>0</td></tr>
<tr><td>1</td><td>0</td><td>0</td></tr>
<tr><td>1</td><td>1</td><td>1</td></tr>
<tr><td>Boolean Expression Q = A ⊕ B</td><td colspan="3">Read if A AND B the SAME gives Q (even)</td></tr>
</table>

Single Input Logic Gates

The Hex Buffer

<table>
<tr><th>Symbol</th><th colspan="2">Truth Table</th></tr>
<tr><td rowspan="3">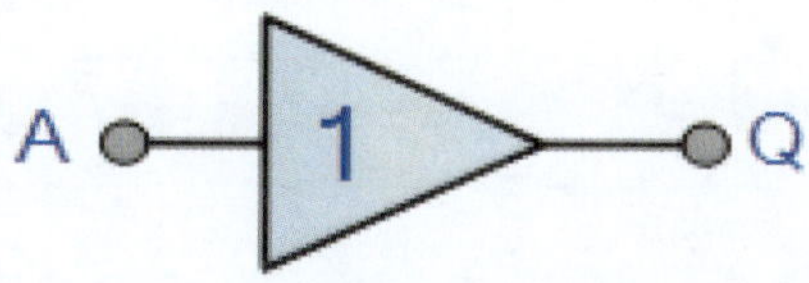
</td><td>A</td><td>Q</td></tr>
<tr><td>0</td><td>0</td></tr>
<tr><td>1</td><td>1</td></tr>
<tr><td>Boolean Expression Q = A</td><td colspan="2">Read as A gives Q</td></tr>
</table>

The NOT Gate (Inverter)

<table>
<tr><th>Symbol</th><th colspan="2">Truth Table</th></tr>
<tr><td rowspan="3">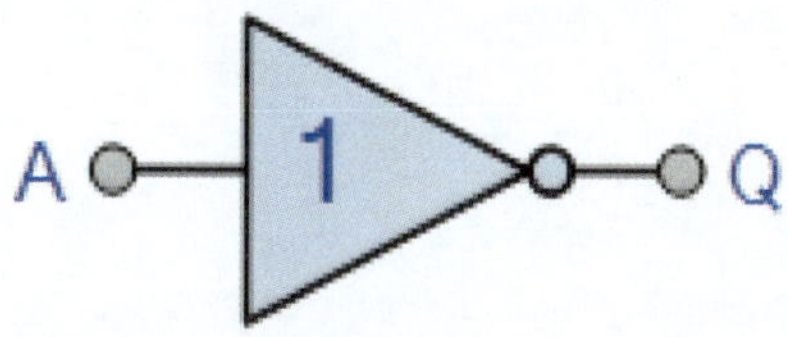
</td><td>A</td><td>Q</td></tr>
<tr><td>0</td><td>1</td></tr>
<tr><td>1</td><td>0</td></tr>
<tr><td>Boolean Expression Q = not A or A</td><td colspan="2">Read as inverse of A gives Q</td></tr>
</table>

The operation of the above Digital Logic Gates and their Boolean expressions can be summarised into a single truth table as shown. This truth table shows the relationship between each output of the main digital logic gates for each possible input combination.

Digital Logic Gate Truth Table

The following logic gates truth table compares the logical functions of the 2-input logic gates detailed.

Inputs		Truth Table Outputs For Each Gate					
B	A	AND	NAND	OR	NOR	EX-OR	EX-NOR
0	0	0	1	0	1	0	1
0	1	0	1	1	0	1	0
1	0	0	1	1	0	1	0
1	1	1	0	1	0	0	1

Truth Table Output for Single-input Gates		
A	NOT	Buffer
0	1	0
1	0	1

Pull-up and Pull-down Resistors

One final point to remember, when connecting together digital logic gates to produce logic circuits, any "unused" inputs to the gates must be connected directly to either a logic level "1" or a logic level "0" by means of a suitable "Pull-up" or "Pull-down" resistor (for example 1kΩ resistor) to produce a fixed logic signal. This will prevent the unused input to the gate from "floating" about and producing false switching of the gate and circuit.

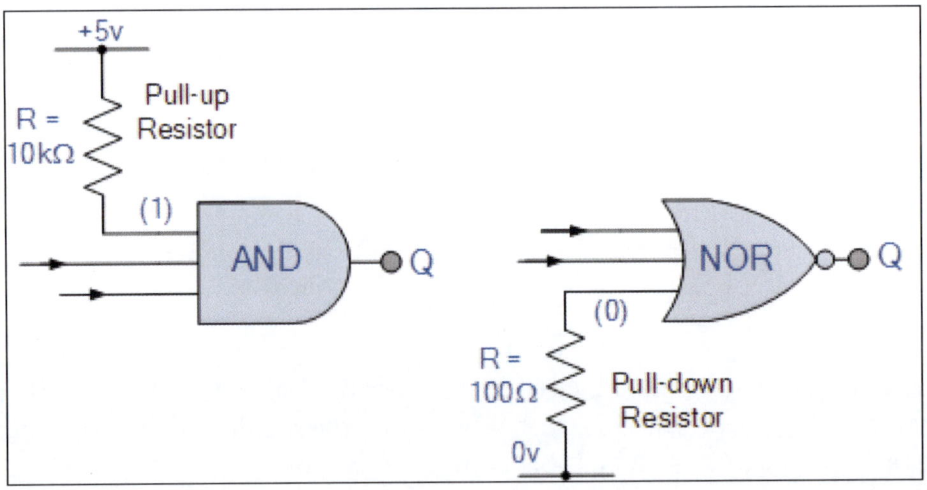

As well as using pull-up or pull-down resistors to prevent unused logic gates from floating about, spare inputs to gates and latches can also be connected together or connected to left-over or spare gates within a single IC package as shown.

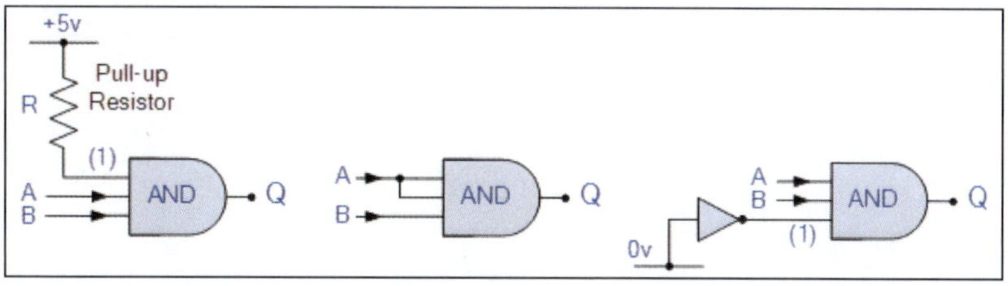

AND Gate

The AND gate is a basic digital logic gate that implements logical conjunction - it behaves according to the truth table to the right. A HIGH output (1) results only if all the inputs to the AND gate are HIGH (1). If none or not all inputs to the AND gate are HIGH, a LOW output results. The function can be extended to any number of inputs.

Symbols

There are three symbols for AND gates: the American (ANSI or 'military') symbol and the IEC ('European' or 'rectangular') symbol, as well as the deprecated DIN symbol. Additional inputs can be added as needed.

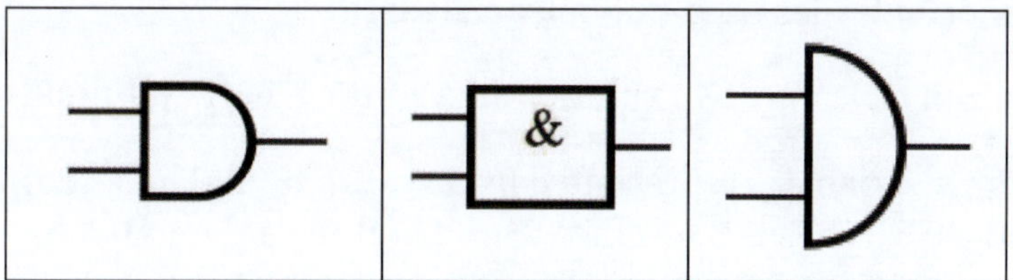

MIL/ANSI Symbol	IEC Symbol	DIN Symbol

The AND gate with inputs *A* and *B* and output *C* implements the logical expression.

Implementations

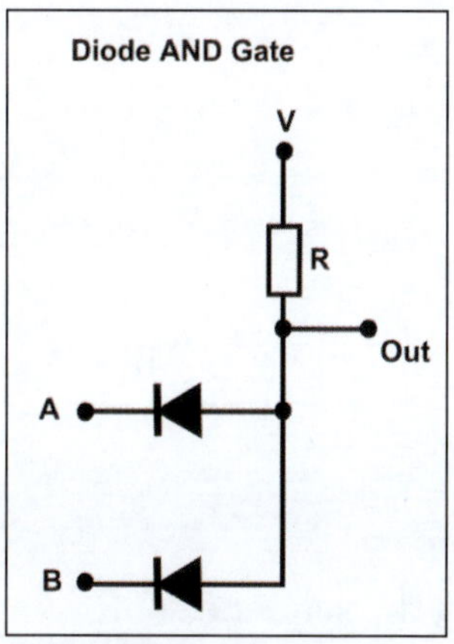

AND gate using diodes.

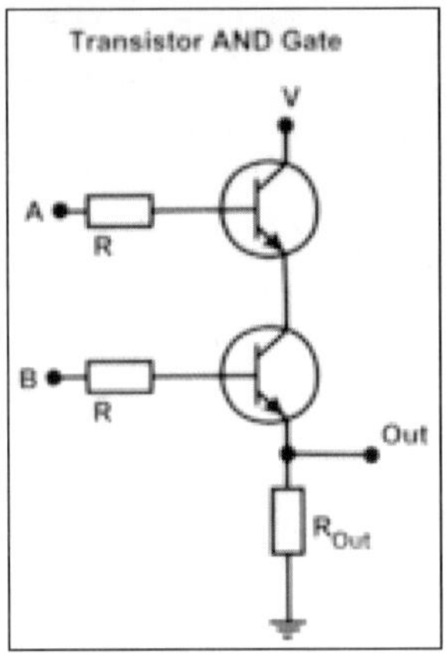

Boolean Algebra

Boolean Algebra is used to analyze and simplify the digital (logic) circuits. It uses only the binary numbers i.e. 0 and 1. It is also called as Binary Algebra or logical Algebra. Boolean algebra was invented by George Boole in 1854.

Rule in Boolean Algebra

Following are the important rules used in Boolean algebra.

- Variable used can have only two values. Binary 1 for HIGH and Binary 0 for LOW.
- Complement of a variable is represented by an overbar (-). Thus, complement of variable B is represented as $\overline{B}$. Thus if B = 0 then $\overline{B}$ = 1 and B = 1 then $\overline{B}$ = 0.
- ORing of the variables is represented by a plus (+) sign between them. For example ORing of A, B, C is represented as A + B + C.
- Logical ANDing of the two or more variable is represented by writing a dot between them such as A.B.C. Sometime the dot may be omitted like ABC.

Boolean Laws

There are six types of Boolean Laws.

Commutative Law

Any binary operation which satisfies the following expression is referred to as commutative operation.

(i) A.B=B.A (ii) A+B=B+A

Commutative law states that changing the sequence of the variables does not have any effect on the output of a logic circuit.

Associative Law

This law states that the order in which the logic operations are performed is irrelevant as their effect is the same.

(i) $(A.B.C = A.(B.C)$ (ii) $(A+B)+C = A+(B+C)$

AND Law

These laws use the AND operation. Therefore they are called as AND laws.

(i) $A+0=A$ (ii) $A+1=1$

(iii) $A+A=A$ (iv) $A+\overline{A}=1$

Inversion Law

This law uses the NOT operation. The inversion law states that double inversion of a variable results in the original variable itself.

$\overline{\overline{A}}=A$

Important Boolean Theorems

Following are few important boolean Theorems.

Boolean function/theorems	Description
Boolean Functions	Boolean Functions and Expressions, K-Map and NAND Gates realization
De Morgan's Theorems	De Morgan's Theorem 1 and Theorem 2

De Morgan's Theorem

In propositional logic and boolean algebra, De Morgan's laws are a pair of transformation rules that are both valid rules of inference. They are named after Augustus De Morgan, a 19th-century British mathematician. The rules allow the expression of conjunctions and disjunctions purely in terms of each other via negation.

The rules can be expressed in English as:

- The negation of a disjunction is the conjunction of the negations; and
- The negation of a conjunction is the disjunction of the negations;

or,

- The complement of the union of two sets is the same as the intersection of their complements; and
- The complement of the intersection of two sets is the same as the union of their complements;

or,

not (A or B) = not A and not B; and

not (A and B) = not A or not B.

In set theory and Boolean algebra, these are written formally as,

$$\overline{A \cup B} = \overline{A} \cap \overline{B},$$

$$\overline{A \cap B} = \overline{A} \cup \overline{B},$$

where.

- A and B are sets,
- A is the complement of A,
- ∩ is the intersection, and
- ∪ is the union.

In formal language, the rules are written as,

$$\neg(P \vee Q) \Leftrightarrow (\neg P) \wedge (\neg Q),$$

and,

$$\neg(P \wedge Q) \Leftrightarrow (\neg P) \vee (\neg Q)$$

where,

- P and Q *are propositions,*
- $\neg$ is the negation logic operator (NOT),
- $\wedge$ is the conjunction logic operator (AND),
- $\vee$ is the disjunction logic operator (OR),
- $\Leftrightarrow$ is a metalogical symbol meaning “can be replaced in a logical proof with”.

Applications of the rules include simplification of logical expressions in computer programs and digital circuit designs. De Morgan’s laws are an example of a more general concept of mathematical duality.

Formal Notation

The *negation of conjunction* rule may be written in sequent notation:

$$\neg(P \wedge Q) \vdash (\neg P \vee \neg Q).$$

The *negation of disjunction* rule may be written as:

$$\neg(P \vee Q) \vdash (\neg P \wedge \neg Q).$$

In rule form: *negation of conjunction,*

$$\frac{\neg(P \wedge Q)}{\therefore \neg P \vee \neg Q}$$

and *negation of disjunction,*

$$\frac{\neg(P \vee Q)}{\therefore \neg P \wedge \neg Q}$$

and expressed as a truth-functional tautology or theorem of propositional logic:

$$\neg(P \wedge Q) \rightarrow (\neg P \vee \neg Q),$$
$$\neg(P \vee Q) \rightarrow (\neg P \wedge \neg Q),$$

where P and Q are propositions expressed in some formal system.

Substitution Form

De Morgan's laws are normally shown in the compact form above, with negation of the output on the left and negation of the inputs on the right. A clearer form for substitution can be stated as:

$$(P \wedge Q) \equiv \neg(\neg P \vee \neg Q),$$
$$(P \vee Q) \equiv \neg(\neg P \wedge \neg Q).$$

This emphasizes the need to invert both the inputs and the output, as well as change the operator, when doing a substitution.

Set theory and Boolean Algebra

In set theory and Boolean algebra, it is often stated as "union and intersection interchange under complementation", which can be formally expressed as:

$$\overline{A \cup B} = \bar{A} \cap \bar{B},$$
$$\overline{A \cap B} = \bar{A} \cup \bar{B},$$

where:

- $\bar{A}$ is the negation of A, the overline being written above the terms to be negated,
- ∩ is the intersection operator (AND),
- ∪ is the union operator (OR).

The generalized form is:

$$\overline{\bigcap_{i \in I} A_i} \equiv \bigcup_{i \in I} \overline{A_i},$$
$$\overline{\bigcup_{i \in I} A_i} \equiv \bigcap_{i \in I} \overline{A_i},$$

where I is some, possibly uncountable, indexing set.

In set notation, De Morgan's laws can be remembered using the mnemonic "break the line, change the sign".

Engineering

In electrical and computer engineering, De Morgan's laws are commonly written as:

$$\overline{A \cdot B} \equiv \overline{A} + \overline{B}$$

and,

$$\overline{A + B} \equiv \overline{A} \cdot \overline{B},$$

where:

- $\cdot$ is the logical AND,
- $+$ is the logical OR,
- The overbar is the logical NOT of what is underneath the $\overline{\text{overbar}}$.

Text Searching

De Morgan's laws commonly apply to text searching using Boolean operators AND, OR, and NOT. Consider a set of documents containing the words "cars" and "trucks". De Morgan's laws hold that these two searches will return the same set of documents:

- Search A: NOT (cars OR trucks);
- Search B: (NOT cars) AND (NOT trucks).

The corpus of documents containing "cars" or "trucks" can be represented by four documents:

- Document 1: Contains only the word "cars".
- Document 2: Contains only "trucks".
- Document 3: Contains both "cars" and "trucks".
- Document 4: Contains neither "cars" nor "trucks".

To evaluate Search A, clearly the search "(cars OR trucks)" will hit on Documents 1, 2, and 3. So the negation of that search (which is Search A) will hit everything else, which is Document 4.

Evaluating Search B, the search "(NOT cars)" will hit on documents that do not contain "cars", which is Documents 2 and 4. Similarly the search "(NOT trucks)" will hit on Documents 1 and 4. Applying the AND operator to these two searches (which is Search B) will hit on the documents that are common to these two searches, which is Document 4.

A similar evaluation can be applied to show that the following two searches will return the same set of documents (Documents 1, 2, 4):

- Search C: NOT (cars AND trucks),
- Search D: (NOT cars) OR (NOT trucks).

Informal Proof

De Morgan's theorem may be applied to the negation of a disjunction or the negation of a conjunction in all or part of a formula.

Negation of a Disjunction

In the case of its application to a disjunction, consider the following claim: "it is false that either of A or B is true", which is written as:

$$\neg(A \vee B).$$

In that it has been established that *neither* A nor B is true, then it must follow that both A is not true and B is not true, which may be written directly as:

$$(\neg A) \wedge (\neg B).$$

If either A or B *were* true, then the disjunction of A and B would be true, making its negation false. Presented in English, this follows the logic that "since two things are both false, it is also false that either of them is true".

Working in the opposite direction, the second expression asserts that A is false and B is false (or equivalently that "not A" and "not B" are true). Knowing this, a disjunction of A and B must be false also. The negation of said disjunction must thus be true, and the result is identical to the first claim.

Negation of a Conjunction

The application of De Morgan's theorem to a conjunction is very similar to its application to a disjunction both in form and rationale. Consider the following claim: "it is false that A and B are both true", which is written as:

$$\neg(A \wedge B).$$

In order for this claim to be true, either or both of A or B must be false, for if they both were true, then the conjunction of A and B would be true, making its negation false. Thus, one (at least) or more of A and B must be false (or equivalently, one or more of "not A" and "not B" must be true). This may be written directly as,

$$(\neg A) \vee (\neg B).$$

Presented in English, this follows the logic that "since it is false that two things are both true, at least one of them must be false".

Working in the opposite direction again, the second expression asserts that at least one of "not A" and "not B" must be true, or equivalently that at least one of A and B must be false. Since at least one of them must be false, then their conjunction would likewise be false. Negating said conjunction thus results in a true expression, and this expression is identical to the first claim.

Formal Proof

The proof that $(A \cap B)^C = A^C \cup B^C$ is completed in 2 steps by proving both $(A \cap B)^C \subseteq A^C \cup B^C$ and $A^C \cup B^C \subseteq (A \cap B)^C$.

Part 1

Let $x \in (A \cap B)^C$. Then, $x \notin A \cap B$.

Because $A \cap B = \{y \mid y \in A \wedge y \in B\}$, it must be the case that $x \notin A$ or $x \notin B$.

If $x \notin A$, then $x \in A^C \cup B^C$, so .

Similarly, if $x \notin B$, then $x \in B^C$, so $x \in A^C \cup B^C$.

Thus, $\forall x(x \in (A \cap B)^C \rightarrow x \in A^C \cup B^C)$;

that is, $(A \cap B)^C \subseteq A^C \cup B^C$.

Part 2

To prove the reverse direction, let $x \in A^C \cup B^C$, and assume $x \notin (A \cap B)^C$.

Under that assumption, it must be the case that $x \in A \cap B$,

so it follows that $x \in A$ and $x \in B$, and thus $x \notin A^C$ and $x \notin B^C$.

However, that means $x \notin A^C \cup B^C$, in contradiction to the hypothesis that $x \in A^C \cup B^C$,

therefore, the assumption $x \notin (A \cap B)^C$ must not be the case, meaning that $x \in (A \cap B)^C$.

Hence, $\forall x(x \in A^C \cup B^C \rightarrow x \in (A \cap B)^C)$,

that is, $A^C \cup B^C \subseteq (A \cap B)^C$.

If $A^C \cup B^C \subseteq (A \cap B)^C$ *and* $(A \cap B)^C \subseteq A^C \cup B^C$, then $(A \cap B)^C = A^C \cup B^C$; this concludes the proof of De Morgan's law.

The other De Morgan's law, $(A \cup B)^C = A^C \cap B^C$, is proven similarly.

Generalising De Morgan Duality

In extensions of classical propositional logic, the duality still holds (that is, to any logical operator one can always find its dual), since in the presence of the identities governing negation, one may always introduce an operator that is the De Morgan dual of another. This leads to an important property of logics based on classical logic, namely the existence of negation normal forms: any formula is equivalent to another formula where negations only occur applied to the non-logical atoms of the formula. The existence of negation normal forms drives many applications, for example in digital circuit design, where it is used to manipulate the types of logic gates, and in formal logic, where it is needed to find the conjunctive normal form and disjunctive normal form of a formula. Computer programmers use them to simplify or properly negate complicated logical conditions. They are also often useful in computations in elementary probability theory.

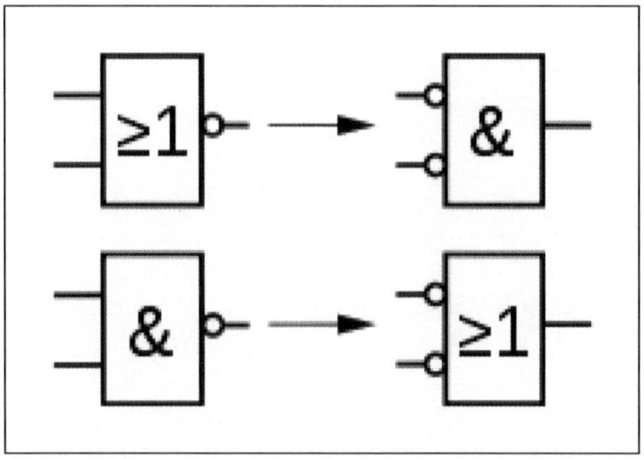

De Morgan's Laws represented as a circuit with logic gates.

Let one define the dual of any propositional operator P(*p*, *q*, ...) depending on elementary propositions *p*, *q*, ... to be the operator P^d defined by,

$$P^d(p,q,...) = \neg P(\neg p, \neg q, \ldots).$$

Extension to Predicate and Modal Logic

This duality can be generalised to quantifiers, so for example the universal quantifier and existential quantifier are duals:

$$\forall x P(x) \equiv \neg[\exists x \neg P(x)]$$

$$\exists x P(x) \equiv \neg[\forall x \neg P(x)]$$

To relate these quantifier dualities to the De Morgan laws, set up a model with some small number of elements in its domain *D*, such as,

$D = \{a, b, c\}$.

Then,

$$\forall x P(x) \equiv P(a) \wedge P(b) \wedge P(c)$$

and,

$$\exists x P(x) \equiv P(a) \vee P(b) \vee P(c).$$

But, using De Morgan's laws,

$$P(a) \wedge P(b) \wedge P(c) \equiv \neg(\neg P(a) \vee \neg P(b) \vee \neg P(c))$$

and,

$$P(a) \vee P(b) \vee P(c) \equiv \neg(\neg P(a) \wedge \neg P(b) \wedge \neg P(c)),$$

verifying the quantifier dualities in the model.

Then, the quantifier dualities can be extended further to modal logic, relating the box ("necessarily") and diamond ("possibly") operators:

$$\Box p \equiv \neg \Diamond \neg p,$$

$$\Diamond p \equiv \neg \Box \neg p.$$

In its application to the alethic modalities of possibility and necessity, Aristotle observed this case, and in the case of normal modal logic, the relationship of these modal operators to the quantification can be understood by setting up models using Kripke semantics.

K-Map

The Karnaugh map (KM or K-map) is a method of simplifying Boolean algebra expressions. Maurice Karnaugh introduced it in 1953 as a refinement of Edward Veitch's 1952 Veitch chart, which actually was a rediscovery of Allan Marquand's 1881 logical diagram aka Marquand diagram but with a focus now set on its utility for switching circuits. Veitch charts are therefore also known as Marquand–Veitch diagrams, and Karnaugh maps as Karnaugh–Veitch maps (KV maps).

The Karnaugh map reduces the need for extensive calculations by taking advantage of humans' pattern-recognition capability. It also permits the rapid identification and elimination of potential race conditions.

The required Boolean results are transferred from a truth table onto a two-dimensional grid where, in Karnaugh maps, the cells are ordered in Gray code, and each cell position represents one combination of input conditions, while each cell value represents the corresponding output value. Optimal groups of 1s or 0s are identified, which represent the terms of a canonical form of the logic in the original truth table. These terms can be used to write a minimal Boolean expression representing the required logic.

Karnaugh maps are used to simplify real-world logic requirements so that they can be implemented using a minimum number of physical logic gates. A sum-of-products expression can always be implemented using AND gates feeding into an OR gate, and a product-of-sums expression leads to OR gates feeding an AND gate. Karnaugh maps can also be used to simplify logic expressions in software design. Boolean conditions, as used for example in conditional statements, can get very complicated, which makes the code difficult to read and to maintain. Once minimised, canonical sum-of-products and product-of-sums expressions can be implemented directly using AND and OR logic operators. Diagrammatic and mechanical methods for minimizing simple logic expressions have existed since at least the medieval times. More systematic methods for minimizing complex expressions began to be developed in the early 1950s, but until the mid to late 1980s the Karnaugh map was the most common used in practice.

Karnaugh maps are used to facilitate the simplification of Boolean algebra functions. For example, consider the Boolean function described by the following truth table.

Truth table of a function					
	A	*B*	*C*	*D*	*f*(A,B,C,D)
0	0	0	0	0	0
1	0	0	0	1	0
2	0	0	1	0	0
3	0	0	1	1	0
4	0	1	0	0	0
5	0	1	0	1	0
6	0	1	1	0	1
7	0	1	1	1	0
8	1	0	0	0	1
9	1	0	0	1	1
10	1	0	1	0	1
11	1	0	1	1	1
12	1	1	0	0	1
13	1	1	0	1	1
14	1	1	1	0	1
15	1	1	1	1	0

Following are two different notations describing the same function in unsimplified Boolean algebra, using the Boolean variables *A*, *B*, *C*, *D*, and their inverses.

- $f(A,B,C,D) = \sum m_i, i \in \{6,8,9,10,11,12,13,14\}$ where m_i are the minterms to map (i.e., rows that have output 1 in the truth table).
- $f(A,B,C,D) = \prod M_i, i \in \{0,1,2,3,4,5,7,15\}$ where M_i are the maxterms to map (i.e., rows that have output 0 in the truth table).

Karnaugh Map

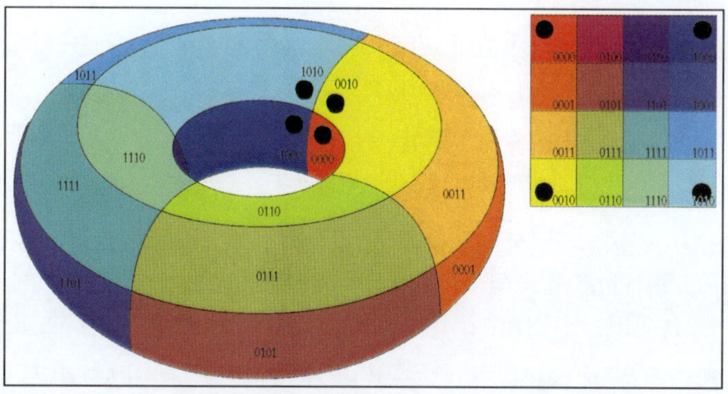

K-map drawn on a torus, and in a plane. The dot-marked cells are adjacent.

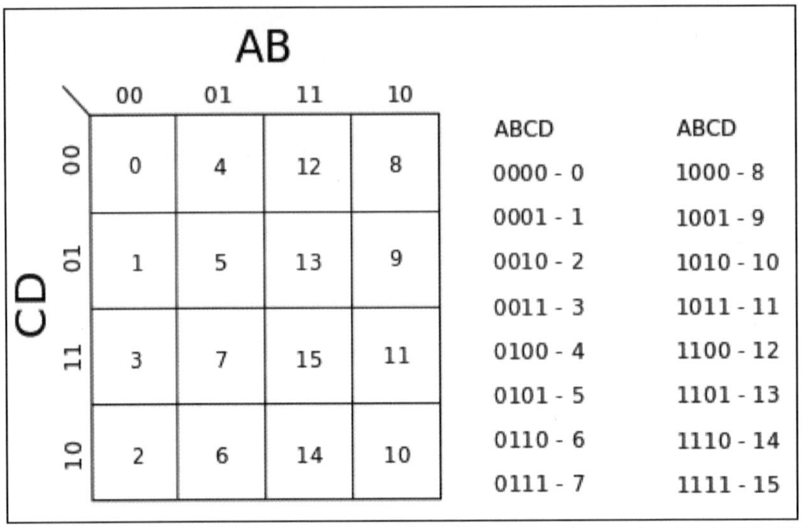

K-map construction. Instead of the output values (the rightmost values in the truth table), this diagram shows a decimal representation of the input ABCD (the leftmost values in the truth table), therefore it is not a Karnaugh map.

In the example above, the four input variables can be combined in 16 different ways, so the truth table has 16 rows, and the Karnaugh map has 16 positions. The Karnaugh map is therefore arranged in a 4 × 4 grid.

The row and column indices (shown across the top, and down the left side of the Karnaugh map) are ordered in Gray code rather than binary numerical order. Gray code ensures that only one variable changes between each pair of adjacent cells. Each cell of the completed Karnaugh map contains a binary digit representing the function's output for that combination of inputs.

After the Karnaugh map has been constructed, it is used to find one of the simplest possible forms – a canonical form – for the information in the truth table. Adjacent 1s in the Karnaugh map represent opportunities to simplify the expression. The minterms ('minimal terms') for the final expression are found by encircling groups of 1s in the map. Minterm groups must be rectangular and must have an area that is a power of two (i.e., 1, 2, 4, 8...). Minterm rectangles should be as large as possible without containing any 0s. Groups may overlap in order to make each one larger. The optimal groupings in the example below are marked by the green, red and blue lines, and the red and green groups overlap. The red group is a 2 × 2 square, the green group is a 4 × 1 rectangle, and the overlap area is indicated in brown.

The cells are often denoted by a shorthand which describes the logical value of the inputs that the cell covers. For example, *AD* would mean a cell which covers the 2x2 area where *A* and *D* are true, i.e. the cells numbered 13, 9, 15, 11 in the diagram above. On the other hand, $A\overline{D}$ would mean the cells where *A* is true and *D* is false (that is, $\overline{D}$ is true).

The grid is toroidally connected, which means that rectangular groups can wrap across the edges. Cells on the extreme right are actually 'adjacent' to those on the far left, in the sense that the corresponding input values only differ by one bit; similarly, so are those at the very top and those at the bottom. Therefore, $A\overline{D}$ can be a valid term—it includes cells 12 and 8 at the top, and wraps to the bottom to include cells 10 and 14—as is $\overline{B}$, $\overline{D}$, which includes the four corners.

Solution:

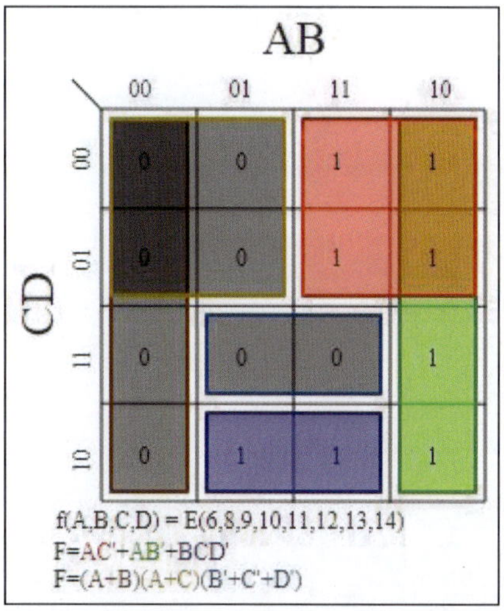

Diagram showing two K-maps. The K-map for the function f(A, B, C, D) is shown as colored rectangles which correspond to minterms. The brown region is an overlap of the red 2×2 square and the green 4×1 rectangle. The K-map for the inverse of f is shown as gray rectangles, which correspond to maxterms.

Once the Karnaugh map has been constructed and the adjacent 1s linked by rectangular and square boxes, the algebraic minterms can be found by examining which variables stay the same within each box.

For the red grouping:

- *A* is the same and is equal to 1 throughout the box, therefore it should be included in the algebraic representation of the red minterm.
- *B* does not maintain the same state (it shifts from 1 to 0), and should therefore be excluded.
- *C* does not change. It is always 0, so its complement, NOT-C, should be included. Thus, $\overline{C}$ should be included.
- *D* changes, so it is excluded.

Thus the first minterm in the Boolean sum-of-products expression is $A\overline{C}$.

For the green grouping, *A* and *B* maintain the same state, while *C* and *D* change. *B* is 0 and has to be negated before it can be included. The second term is therefore $A\overline{B}$. Note that it is acceptable that the green grouping overlaps with the red one.

In the same way, the blue grouping gives the term $BC\overline{D}$.

The solutions of each grouping are combined: the normal form of the circuit is $A\overline{C} + A\overline{B} + BC\overline{D}$.

Thus the Karnaugh map has guided a simplification of,

$$\begin{aligned} f(A,B,C,D) = {} & \overline{A}BC\overline{D} + A\overline{B}\,\overline{C}\,\overline{D} + A\overline{B}\,\overline{C}D + A\overline{B}C\overline{D} + \\ & A\overline{B}CD + AB\overline{C}\,\overline{D} + AB\overline{C}D + ABC\overline{D} \\ = {} & A\overline{C} + A\overline{B} + BC\overline{D} \end{aligned}$$

It would also have been possible to derive this simplification by carefully applying the axioms of boolean algebra, but the time it takes to do that grows exponentially with the number of terms.

Inverse

The inverse of a function is solved in the same way by grouping the 0s instead.

The three terms to cover the inverse are all shown with grey boxes with different colored borders:

- Brown: $\overline{A}$, $\overline{B}$
- Gold: $\overline{A}$, $\overline{C}$
- Blue: BCD

This yields the inverse:

$$\overline{f(A,B,C,D)} = \overline{A}\overline{B} + \overline{A}\overline{C} + BCD$$

Through the use of De Morgan's laws, the product of sums can be determined:

$$\begin{aligned} f(A,B,C,D) &= \overline{\overline{f(A,B,C,D)}} \\ &= \overline{\overline{A}\overline{B} + \overline{A}\overline{C} + BCD} \\ &= \left(\overline{\overline{A}\overline{B}}\right)\left(\overline{\overline{A}\overline{C}}\right)\left(\overline{BCD}\right) \\ &= (A+B)(A+C)\left(\overline{B}+\overline{C}+\overline{D}\right) \end{aligned}$$

'Don't Care' Condition

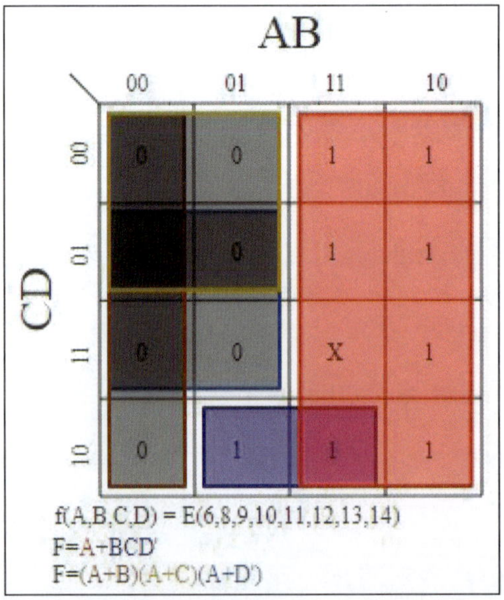

The value of $f(A,B,C,D)$ for $ABCD$ = 1111 is replaced by a "don't care". This removes the green term completely and allows the red term to be larger. It also allows blue inverse term to shift and become larger.

Karnaugh maps also allow easy minimizations of functions whose truth tables include "don't care" conditions. A "don't care" condition is a combination of inputs for which the designer doesn't care what the output is. Therefore, "don't care" conditions can either be included in or excluded from any rectangular group, whichever makes it larger. They are usually indicated on the map with a dash or X.

The example on the right is the same as the example above but with the value of $f(1,1,1,1)$ replaced by a "don't care". This allows the red term to expand all the way down and, thus, removes the green term completely.

This yields the new minimum equation:

$$f(A,B,C,D) = A + BC\overline{D}$$

Note that the first term is just *A*, not *AC*. In this case, the don't care has dropped a term (the green rectangle); simplified another (the red one); and removed the race hazard.

The inverse case is simplified as follows:

$$\overline{f(A,B,C,D)} = \overline{A}\overline{B} + \overline{A}\overline{C} + \overline{A}D$$

Race Hazards

Elimination

Karnaugh maps are useful for detecting and eliminating race conditions. Race hazards are very easy to spot using a Karnaugh map, because a race condition may exist when moving between any pair of adjacent, but disjoint, regions circumscribed on the map. However, because of the nature of Gray coding, *adjacent* has a special definition explained above – we're in fact moving on a torus, rather than a rectangle, wrapping around the top, bottom, and the sides.

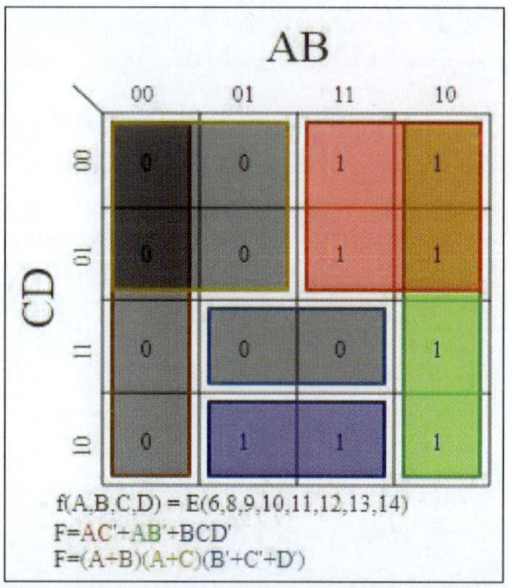

Race hazards are present in this diagram.

- In the example above, a potential race condition exists when *C* is 1 and *D* is 0, *A* is 1, and *B* changes from 1 to 0 (moving from the blue state to the green state). For this case, the output is defined to remain unchanged at 1, but because this transition is not covered by a specific term in the equation, a potential for a *glitch* (a momentary transition of the output to 0) exists.

- There is a second potential glitch in the same example that is more difficult to spot: when D is 0 and A and B are both 1, with C changing from 1 to 0 (moving from the blue state to the red state). In this case the glitch wraps around from the top of the map to the bottom.

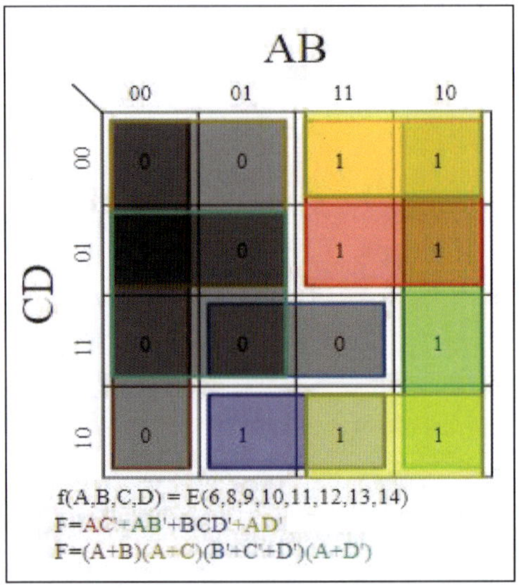

Above diagram with consensus terms added to avoid race hazards.

Whether glitches will actually occur depends on the physical nature of the implementation, and whether we need to worry about it depends on the application. In clocked logic, it is enough that the logic settles on the desired value in time to meet the timing deadline. In our example, we are not considering clocked logic.

In our case, an additional term of $A\overline{D}$ would eliminate the potential race hazard, bridging between the green and blue output states or blue and red output states: this is shown as the yellow region (which wraps around from the bottom to the top of the right half) in the diagram.

The term is redundant in terms of the static logic of the system, but such redundant, or consensus terms, are often needed to assure race-free dynamic performance.

Similarly, an additional term of $\overline{A}D$ must be added to the inverse to eliminate another potential race hazard. Applying De Morgan's laws creates another product of sums expression for *f*, but with a new factor of $\left(A+\overline{D}\right)$.

2-variable Map Examples

The following are all the possible 2-variable, 2 × 2 Karnaugh maps. Listed with each is the minterms as a function of $\sum m()$ and the race hazard free minimum equation. A minterm is defined as an expression that gives the most minimal form of expression of the mapped variables. All possible horizontal and vertical interconnected blocks can be formed. These blocks must be of the size of the powers of 2 (1, 2, 4, 8, 16, 32, ...). These expressions create a minimal logical mapping of the minimal logic variable expressions for the binary expressions to be mapped. Here are all the blocks with one field.

A block can be continued across the bottom, top, left, or right of the chart. That can even wrap beyond the edge of the chart for variable minimization. This is because each logic variable corresponds to each vertical column and horizontal row. A visualization of the k-map can be considered cylindrical. The fields at edges on the left and right are adjacent, and the top and bottom are adjacent. K-Maps for four variables must be depicted as a donut or torus shape. The four corners of the square drawn by the k-map are adjacent. Still more complex maps are needed for 5 variables and more.

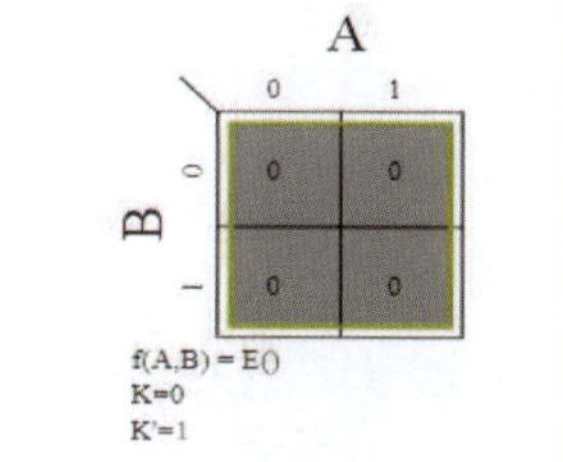	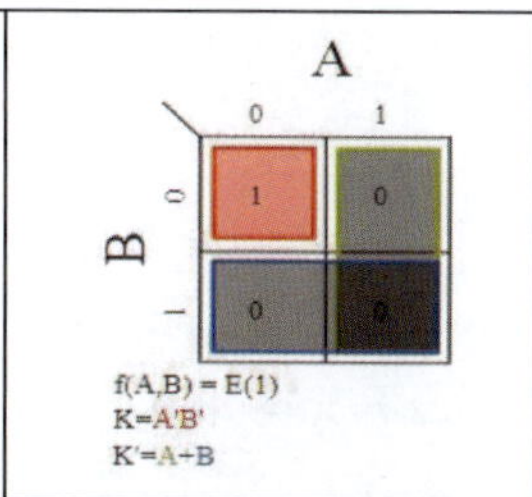	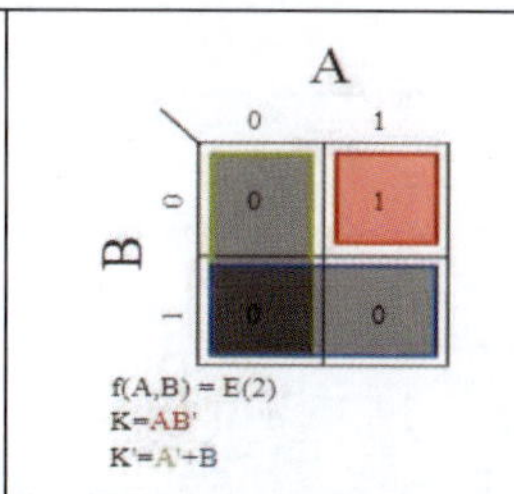	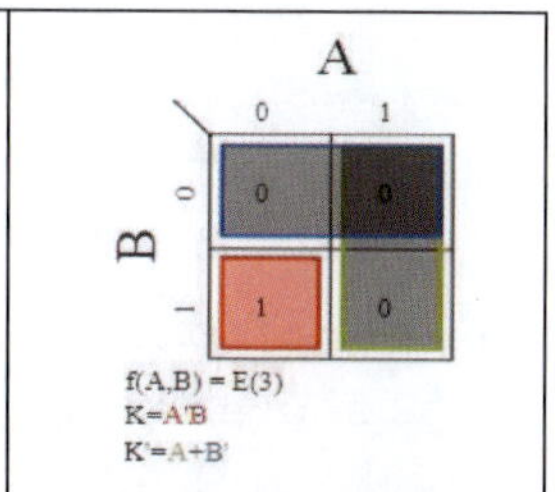
$\Sigma m(0); K = 0$	$\Sigma m(1); K = A'B'$	$\Sigma m(2); K = AB'$	$\Sigma m(3); K = A'B$
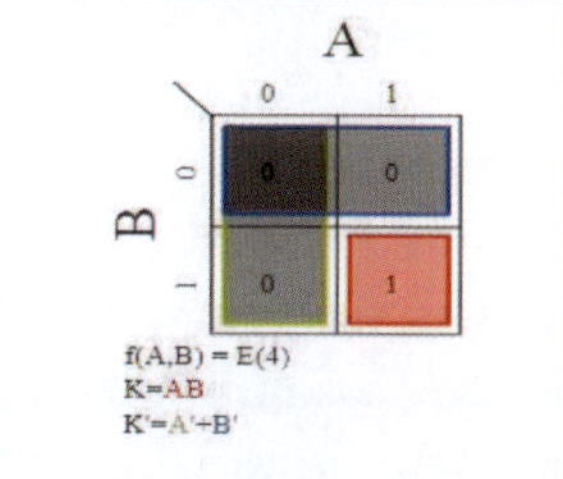	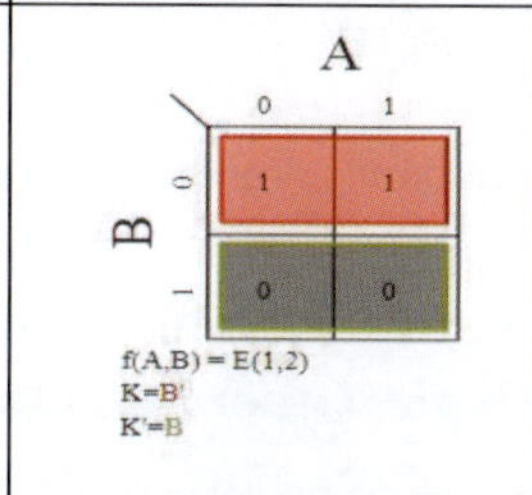	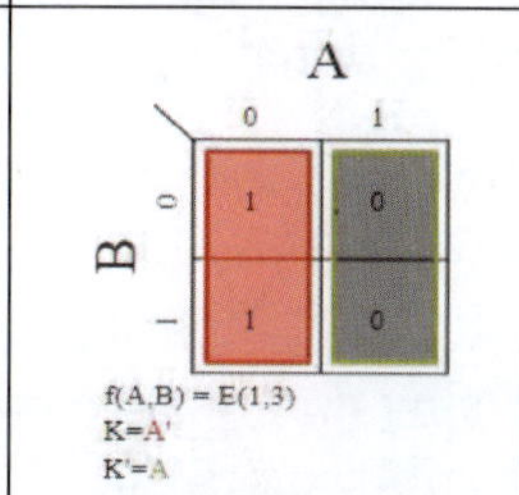	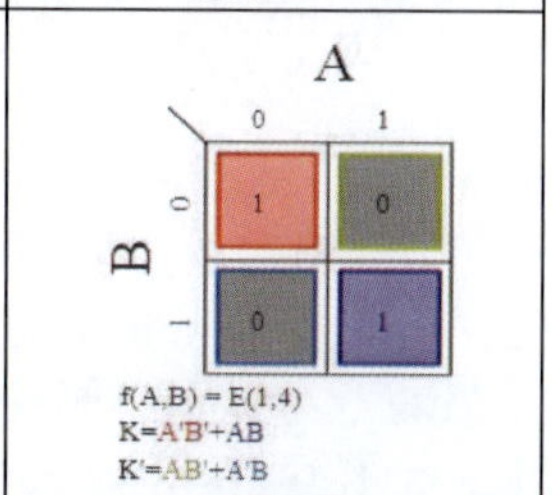
$\Sigma m(4); K = AB$	$\Sigma m(1,2); K = B'$	$\Sigma m(1,3); K = A'$	$\Sigma m(1,4); K = A'B' + AB$
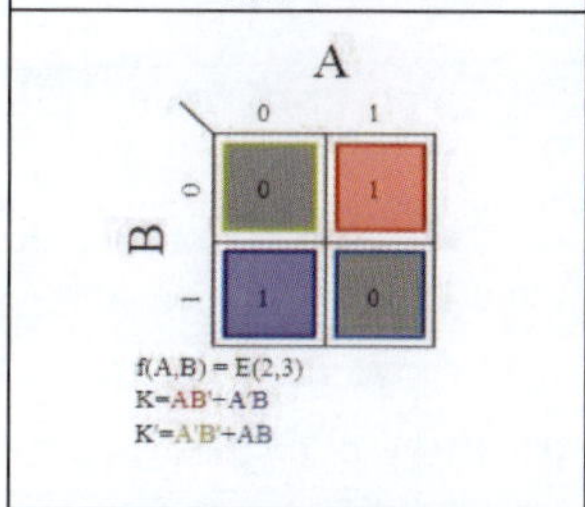	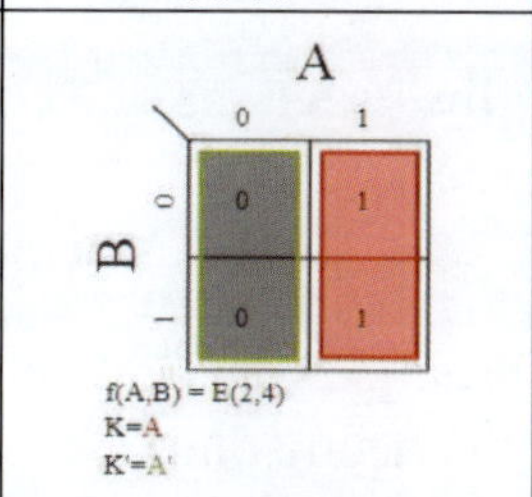	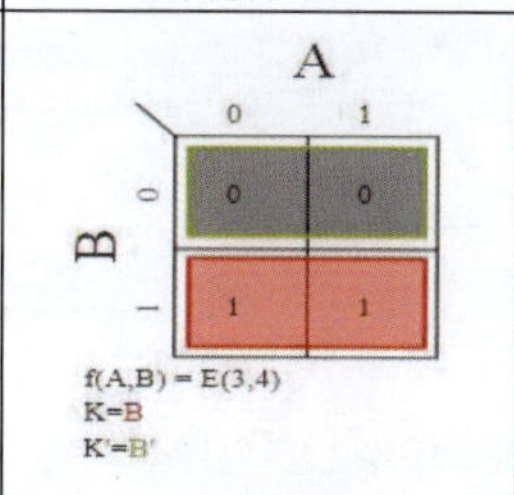	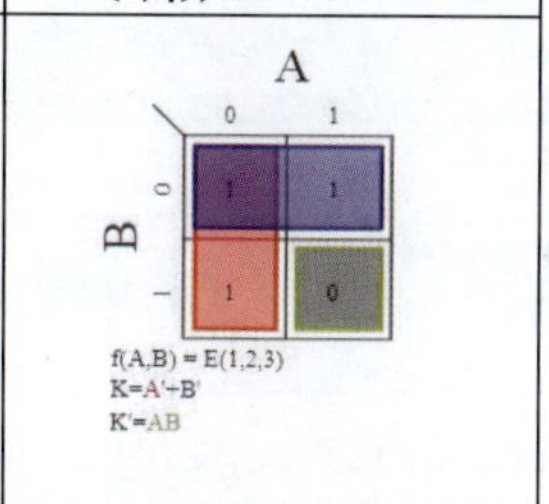
$\Sigma m(2,3); K = AB' + A'B$	$\Sigma m(2,4); K = A$	$\Sigma m(3,4); K = B$	$\Sigma m(1,2,3); K = A' + B'$
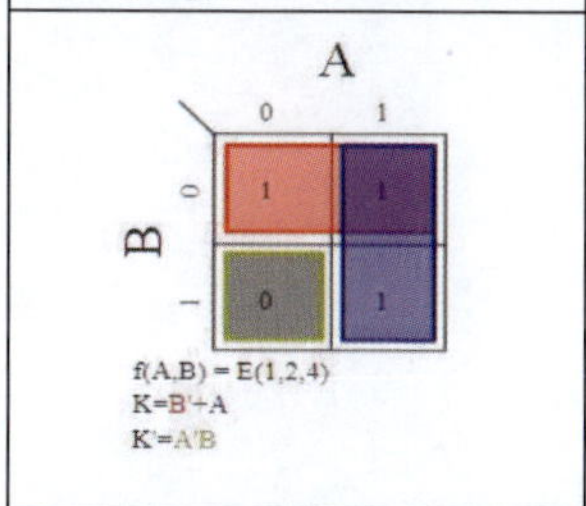	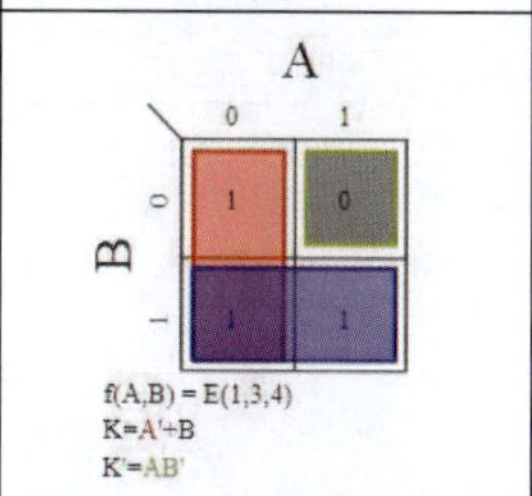	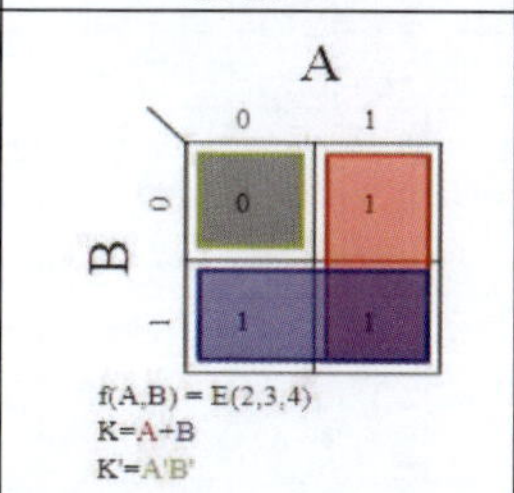	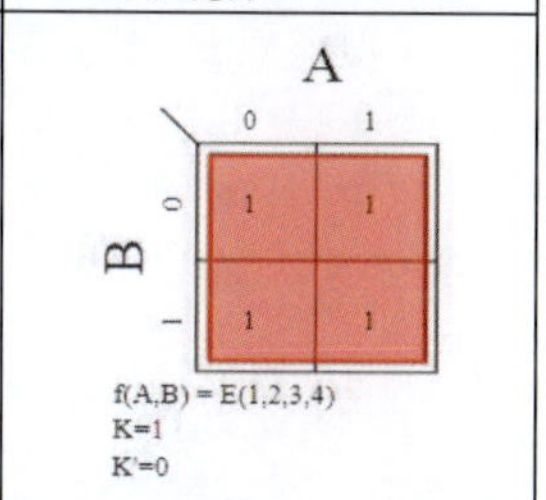
$\Sigma m(1,2,4); K = A + B'$	$\Sigma m(1,3,4); K = A' + B$	$\Sigma m(2,3,4); K = A + B$	$\Sigma m(1,2,3,4); K = 1$

Other Graphical Methods

Alternative graphical minimization methods include:

- Marquand diagram by Allan Marquand;
- *Harvard minimizing chart* by Howard H. Aiken and Martha L. Whitehouse of the Harvard Computation Laboratory;

- *Veitch chart* by Edward Veitch;
- Svoboda's graphical aids and *triadic map* by Antonín Svoboda;
- *Händler circle graph* by Wolfgang Händler;
- Graph method by Herbert Kortum [de].

Combinational Logic

In digital circuit theory, combinational logic (sometimes also referred to as time-independent logic) is a type of digital logic which is implemented by Boolean circuits, where the output is a pure function of the present input only. This is in contrast to sequential logic, in which the output depends not only on the present input but also on the history of the input. In other words, sequential logic has *memory* while combinational logic does not.

Combinational logic is used in computer circuits to perform Boolean algebra on input signals and on stored data. Practical computer circuits normally contain a mixture of combinational and sequential logic. For example, the part of an arithmetic logic unit, or ALU, that does mathematical calculations is constructed using combinational logic. Other circuits used in computers, such as half adders, full adders, half subtractors, full subtractors, multiplexers, demultiplexers, encoders and decoders are also made by using combinational logic.

An alternate term is combinatorial logic, though this usage may be considered controversial.

Representation

Combinational logic is used to build circuits that produce specified outputs from certain inputs. The construction of combinational logic is generally done using one of two methods: a sum of products, or a product of sums. Consider the following truth table (input is considered to be truthful):

A	B	C	Result	Logical equivalent
F	F	F	F	$\neg A. \neg B. \neg C$
F	F	T	F	$\neg A. \neg B. C$
F	T	F	F	$\neg A. B. \neg C$
F	T	T	F	$\neg A. B. C$
T	F	F	T	$A. \neg B. \neg C$
T	F	T	F	$A. \neg B. C$
T	T	F	F	$A. B. \neg C$
T	T	T	T	$A. B. C$

Using sum of products, all logical statements which yield true results are summed, giving the result:

$$A \cdot \neg B \cdot \neg C + A \cdot B \cdot C$$

Using Boolean algebra, the result simplifies to the following equivalent of the truth table:

$$A \cdot (\neg B \cdot \neg C + B \cdot C)$$

Logic Formula Minimization

Minimization (simplification) of combinational logic formulas is done using the following rules based on the laws of Boolean algebra:

$$(A+B)\cdot(A+C) = A+(B\cdot C)$$
$$(A\cdot B)+(A\cdot C) = A\cdot(B+C)$$

$$A+(A\cdot B) = A$$
$$A\cdot(A+B) = A$$

$$A+(\neg A\cdot B) = A+B$$
$$A\cdot(\neg A+B) = A\cdot B$$

$$(A+B)\cdot(\neg A+B) = B$$
$$(A\cdot B)+(\neg A\cdot B) = B$$

$$(A\cdot B)+(\neg A\cdot C)+(B\cdot C) = (A\cdot B)+(\neg A\cdot C)$$
$$(A+B)\cdot(\neg A+C)\cdot(B+C) = (A+B)\cdot(\neg A+C)$$

With the use of minimization (sometimes called logic optimization), a simplified logical function or circuit may be arrived upon, and the logic combinational circuit becomes smaller, and easier to analyse, use, or build.

Multiplexer

In electronics, a multiplexer (or mux) is a device that selects between several analog or digital input signals and forwards it to a single output line. A multiplexer of 2^n inputs has select lines, which are used to select which input line to send to the output. Multiplexers are mainly used to increase the amount of data that can be sent over the network within a certain amount of time and bandwidth. A multiplexer is also called a data selector. Multiplexers can also be used to implement Boolean functions of multiple variables.

An electronic multiplexer makes it possible for several signals to share one device or resource, for example, one A/D converter or one communication line, instead of having one device per input signal.

Conversely, a demultiplexer (or demux) is a device taking a single input and selecting signals of the output of the compatible mux, which is connected to the single input, and a shared selection line. A multiplexer is often used with a complementary demultiplexer on the receiving end.

An electronic multiplexer can be considered as a multiple-input, single-output switch, and a demultiplexer as a single-input, multiple-output switch. The schematic symbol for a multiplexer is an isosceles trapezoid with the longer parallel side containing the input pins and the short parallel side containing the output pin. The sel wire connects the desired input to the output.

Cost Saving

One use for multiplexers is economizing connections over a single channel, by connecting the multiplexer's single output to the demultiplexer's single input. The image to the right demonstrates this benefit. In this case, the cost of implementing separate channels for each data source is higher than the cost and inconvenience of providing the multiplexing/demultiplexing functions.

At the receiving end of the data link a complementary *demultiplexer* is usually required to break the single data stream back down into the original streams. In some cases, the far end system may have functionality greater than a simple demultiplexer; and while the demultiplexing still occurs technically, it may never be implemented discretely. This would be typical when: a multiplexer serves a number of IP network users; and then feeds directly into a router, which immediately reads the content of the entire link into its routing processor; and then does the demultiplexing in memory from where it will be converted directly into IP sections.

Often, a multiplexer and demultiplexer are combined together into a single piece of equipment, which is conveniently referred to as a "multiplexer". Both circuit elements are needed at both ends of a transmission link because most communications systems transmit in both directions.

In analog circuit design, a multiplexer is a special type of analog switch that connects one signal selected from several inputs to a single output.

Digital Multiplexers

In digital circuit design, the selector wires are of digital value. In the case of a 2-to-1 multiplexer, a logic value of 0 would connect I_0 to the output while a logic value of 1 would connect I_1 to the output. In larger multiplexers, the number of selector pins is equal to $\lceil \log_2(n) \rceil$ where n is the number of inputs.

For example, 9 to 16 inputs would require no fewer than 4 selector pins and 17 to 32 inputs would require no fewer than 5 selector pins. The binary value expressed on these selector pins determines the selected input pin.

A 2-to-1 multiplexer has a boolean equation where A and B are the two inputs, S_0 is the selector input, and Z is the output:

$$Z = (A \wedge \neg S_0) \vee (B \wedge S)$$

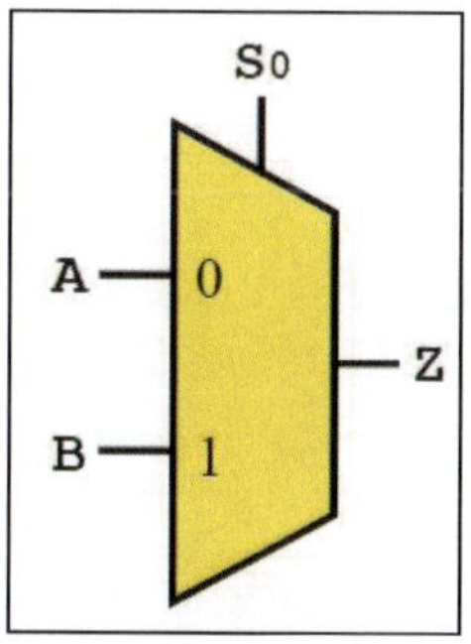

A 2-to-1 mux.

Which can be expressed as a truth table:

S_o	A	B	Z
0	0	0	0
0	0	1	0
0	1	0	1
0	1	1	1
1	0	0	0
1	0	1	1
1	1	0	0
1	1	1	1

Or, in simpler notation:

S_0	Z
0	A
1	B

These tables show that when $S_0 = 0$ then $Z = A$ but when $S_0 = 1$ then $Z = B$. A straightforward realization of this 2-to-1 multiplexer would need 2 AND gates, an OR gate, and a NOT gate. While this is mathematically correct, a direct physical implementation would be prone to race conditions that require additional gates to suppress.

Larger multiplexers are also common and, as stated above, require $\lceil \log_2(n) \rceil$ selector pins for n inputs. Other common sizes are 4-to-1, 8-to-1, and 16-to-1. Since digital logic uses binary values, powers of 2 are used (4, 8, 16) to maximally control a number of inputs for the given number of selector inputs.

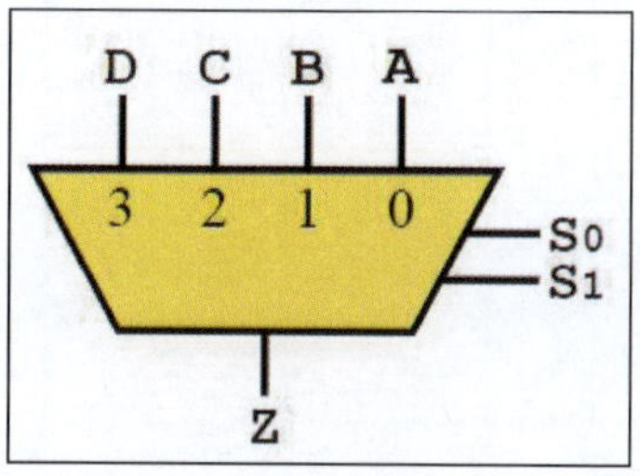

4-to-1 mux.

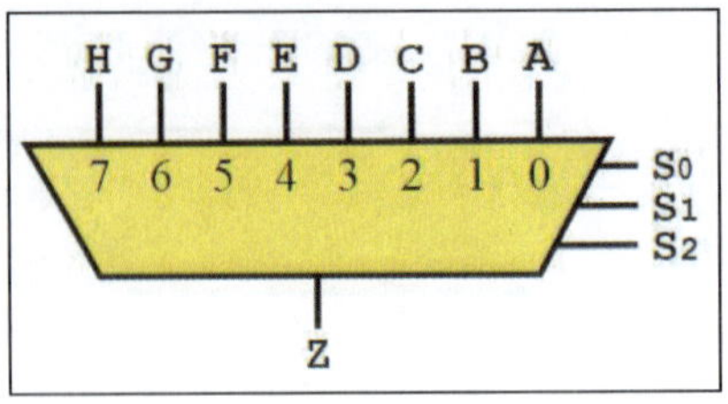

8-to-1 mux.

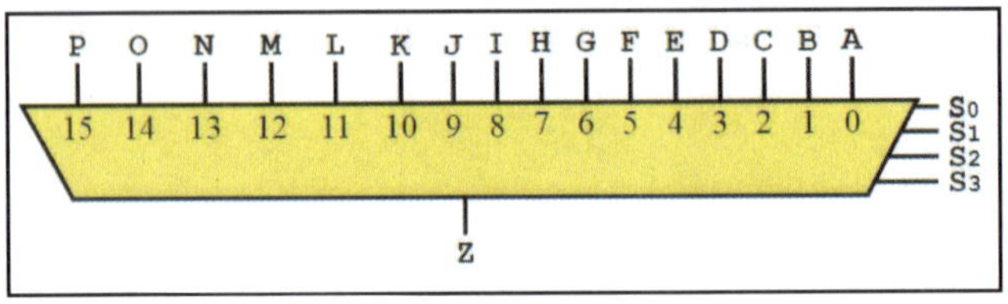

16-to-1 mux.

The boolean equation for a 4-to-1 multiplexer is:

$$Z = (A \wedge \neg S_0 \wedge \neg S_1) \vee (B \wedge S_0 \wedge \neg S_1) \vee (C \wedge \neg S_0 \wedge S_1) \vee (D \wedge S_0 \wedge S_1)$$

The following 4-to-1 multiplexer is constructed from 3-state buffers and AND gates (the AND gates are acting as the decoder):

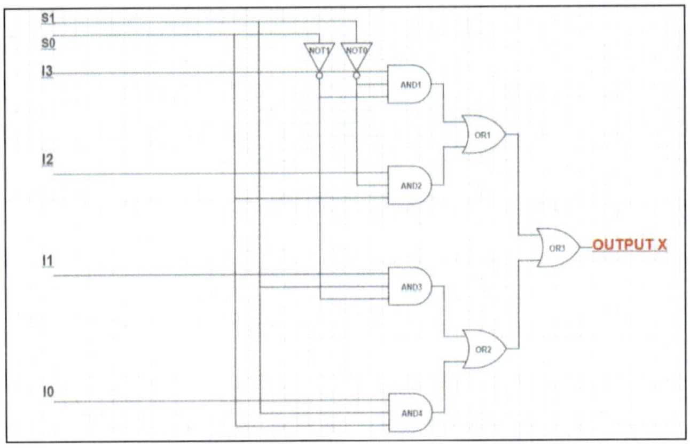

A 4:1 MUX circuit using 3 input AND and other gates.

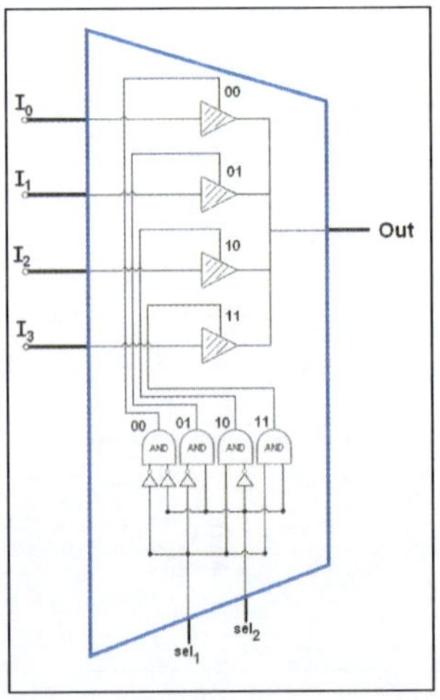

The subscripts on the I_n inputs indicate the decimal value of the binary control inputs at which that input is let through.

Chaining Multiplexers

Larger Multiplexers can be constructed by using smaller multiplexers by chaining them together. For example, an 8-to-1 multiplexer can be made with two 4-to-1 and one 2-to-1 multiplexers. The two 4-to-1 multiplexer outputs are fed into the 2-to-1 with the selector pins on the 4-to-1's put in parallel giving a total number of selector inputs to 3, which is equivalent to an 8-to-1.

List of ICs which Provide Multiplexing

Signetics S54S157.

The 7400 series has several ICs that contain multiplexer(s):

IC No.	Function	Output State
74157	Quad 2:1 mux.	Output same as input given
74158	Quad 2:1 mux.	Output is inverted input
74153	Dual 4:1 mux.	Output same as input
74352	Dual 4:1 mux.	Output is inverted input
74151A	8:1 mux.	Both outputs available (i.e., complementary outputs)
74151	8:1 mux.	Output is inverted input
74150	16:1 mux.	Output is inverted input

Digital Demultiplexers

Demultiplexers take one data input and a number of selection inputs, and they have several outputs. They forward the data input to one of the outputs depending on the values of the selection inputs. Demultiplexers are sometimes convenient for designing general purpose logic, because if the demultiplexer's input is always true, the demultiplexer acts as a decoder. This means that any function of the selection bits can be constructed by logically OR-ing the correct set of outputs.

If X is the input and S is the selector, and A and B are the outputs:

$$A = (X \wedge \neg S)$$

$$B = (X \wedge S)$$

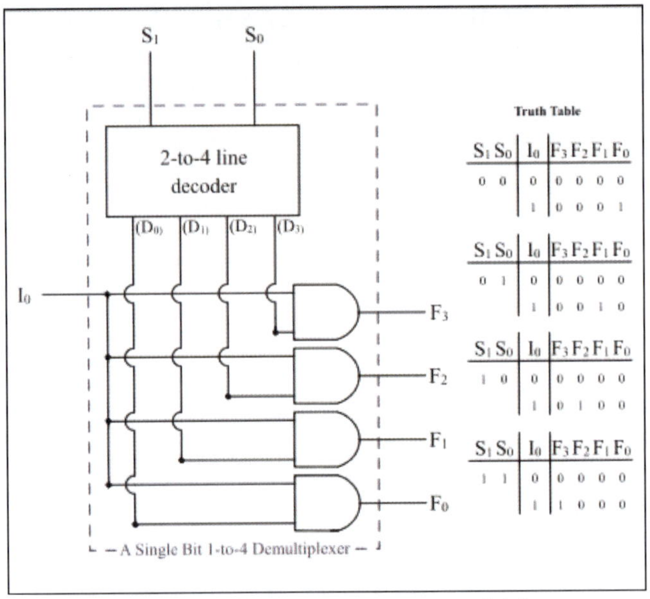

Example: A Single Bit 1-to-4 Line Demultiplexer.

List of ICs which Provide Demultiplexing

Fairchild 74F138.

The 7400 series has several ICs that contain demultiplexer(s):

IC No. (7400)	IC No. (4000)	Function	Output State
74139		Dual 1:4 demux.	Output is inverted input
74156		Dual 1:4 demux.	Output is open collector
74138		1:8 demux.	Output is inverted input
74238		1:8 demux.	
74154		1:16 demux.	Output is inverted input
74159	CD4514/15	1:16 demux.	Output is open collector and same as input

Multiplexers as PLDs

Multiplexers can also be used as programmable logic devices, specifically to implement Boolean functions. Any Boolean function of *n* variables and one result can be implemented with a multiplexer with *n* selector inputs. The variables are connected to the selector inputs, and the function result, 0 or 1, for each possible combination of selector inputs is connected to the corresponding data input. This is especially useful in situations when cost is a factor, for modularity, and for ease of modification. If one of the variables (for example, *D*) is also available inverted, a multiplexer with *n*-1 selector inputs is sufficient; the data inputs are connected to 0, 1, *D*, or ~*D*, according to the desired output for each combination of the selector inputs.

Demultiplexer

The demultiplexer is a combinational logic circuit designed to switch one common input line to one of several seperate output line.

The data distributor, known more commonly as a Demultiplexer or "Demux" for short, is the exact opposite of the Multiplexer.

The demultiplexer takes one single input data line and then switches it to any one of a number of individual output lines one at a time. The demultiplexer converts a serial data signal at the input to a parallel data at its output lines as shown below.

1-to-4 Channel De-multiplexer

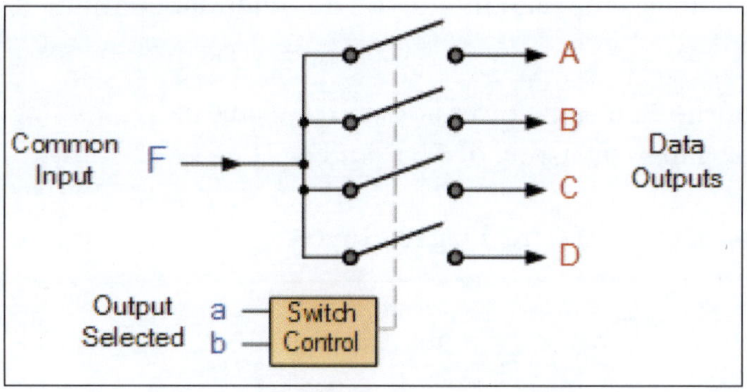

Output Select		Data Output Selected
a	b	
0	0	A
0	1	B
1	0	C
1	1	D

The Boolean expression for this 1-to-4 Demultiplexer above with outputs A to D and data select lines a, b is given as:

$$F = \overline{ab}A + \overline{a}bB + a\overline{b}C + abD$$

The function of the Demultiplexer is to switch one common data input line to any one of the 4 output data lines A to D in our example above. As with the multiplexer the individual solid state switches are selected by the binary input address code on the output select pins "a" and "b" as shown.

Demultiplexer Output Line Selection

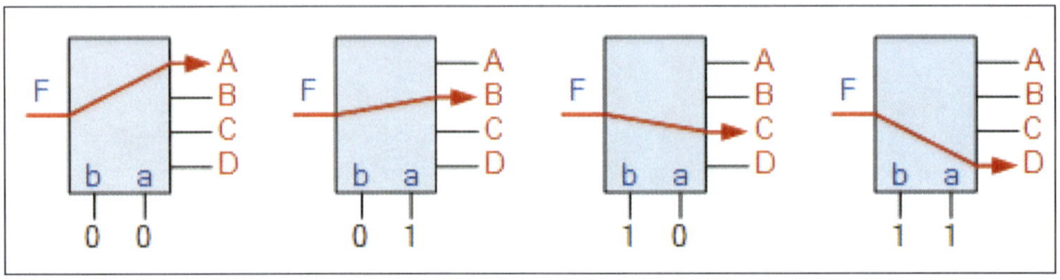

As with the previous multiplexer circuit, adding more address line inputs it is possible to switch more outputs giving a 1-to-2^n data line outputs.

Some standard demultiplexer IC´s also have an additional "enable output" pin which disables or prevents the input from being passed to the selected output. Also some have latches built into their outputs to maintain the output logic level after the address inputs have been changed.

However, in standard decoder type circuits the address input will determine which single data output will have the same value as the data input with all other data outputs having the value of logic "0".

The implementation of the Boolean expression above using individual logic gates would require the use of six individual gates consisting of AND and NOT gates as shown.

4 Channel Demultiplexer using Logic Gates

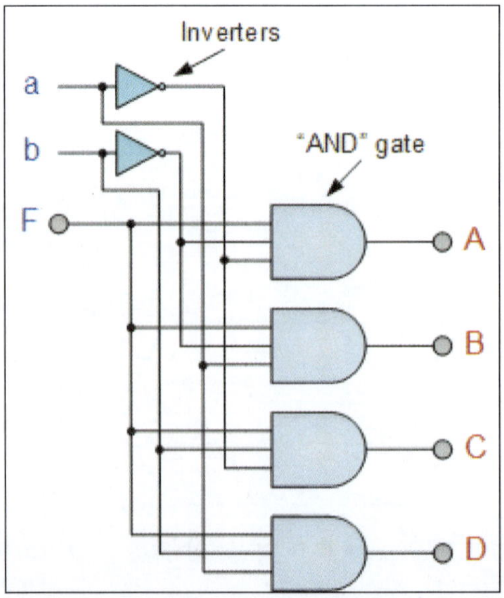

The symbol used in logic diagrams to identify a demultiplexer is as follows.

The Demultiplexer Symbol

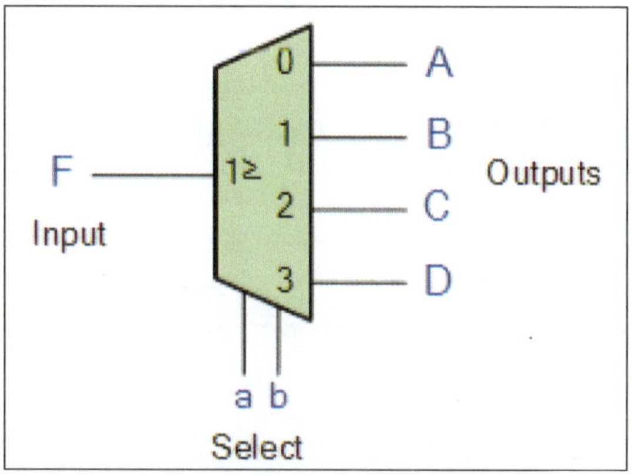

Again, as with the previous multiplexer example, we can also use the demultiplexer to digitally control the gain of an operational amplifier as shown.

Digitally Adjustable Amplifier Gain

The circuit above illustrates how to provide digitally controlled adjustable/variable op-amp gain using a demultiplexer. The voltage gain of the inverting operational amplifier is dependent upon the ratio between the input resistor, R_{IN} and its feedback resistor, Rf as determined.

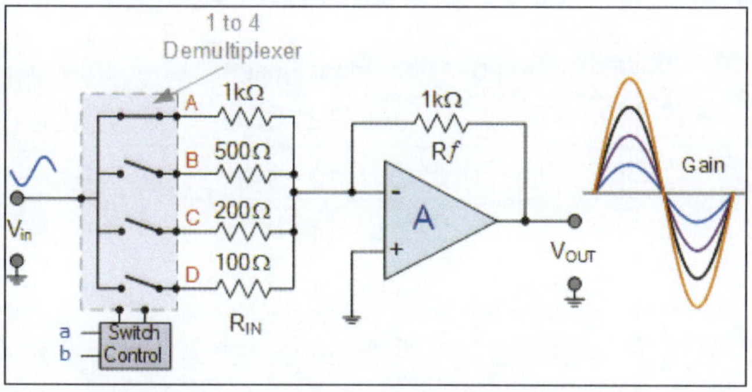

The digitally controlled analogue switches of the demultiplexer select an input resistor to vary the value of Rin. The combination of these resistors will determine the overall voltage gain of the amplifier, (Av). Then the voltage gain of the inverting operational amplifier can be adjusted digitally simply by selecting the appropriate input resistor combination.

Standard Demultiplexer IC packages available are the TTL 74LS138 1 to 8-output demultiplexer, the TTL 74LS139 Dual 1-to-4 output demultiplexer or the CMOS CD4514 1-to-16 output demultiplexer.

Another type of demultiplexer is the 24-pin, 74LS154 which is a 4-bit to 16-line demultiplexer/decoder. Here the individual output positions are selected using a 4-bit binary coded input. Like multiplexers, demultiplexers can also be cascaded together to form higher order demultiplexers.

Unlike multiplexers which convert data from a single data line to multiple lines and demultiplexers which convert multiple lines to a single data line, there are devices available which convert data to and from multiple lines.

Encoder

A simple encoder or simply an encoder in digital electronics is a one-hot to binary converter. That is, if there are 2^n input lines, and at most only one of them will ever be high, the binary code of this 'hot' line is produced on the n-bit output lines.

For example, a 4-to-2 simple encoder takes 4 input bits and produces 2 output bits. The illustrated gate level example implements the simple encoder defined by the truth table, but it must be understood that for all the non-explicitly defined input combinations (i.e., inputs containing 0, 2, 3, or 4 high bits) the outputs are treated as don't cares.

It is the reverse of a Decoder in its function.

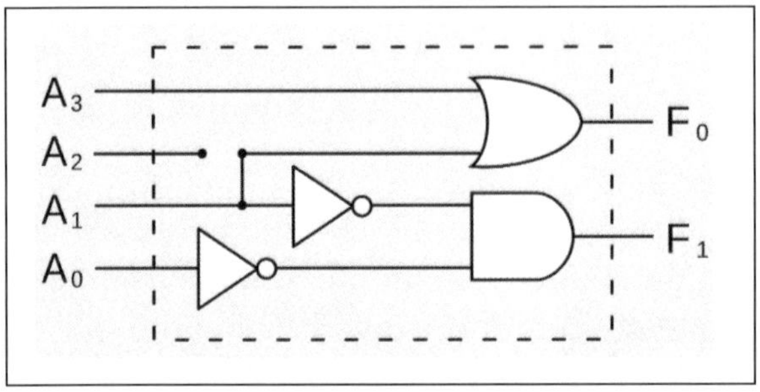

Gate level circuit diagram of a single bit 4-to-2 line encoder.

4 to 2 Simple Encoder						
I_3	I_2	I_1	I_0	O_1	O_0	V
0	0	0	0	X	X	0
0	0	0	1	0	0	1
0	0	1	0	0	1	1
0	1	0	0	1	0	1
1	0	0	0	1	1	1

If the input circuit can guarantee at most a single-active input, a simple encoder is a better choice than a priority encoder, since it requires less logic to implement but if not this become the major disadvantage of simple encoder that is they can generate the wrong output when there is more than one input present in a high state (logic state "1"). To rectify this problem a Priority encoder is used. They act on the request of higher priority and the rest cases go in "don't care condition".

Types of Encoder

-to-n encoders:

A 2^n -to-n encoder has n number of outputs in correspondence to the 2^n number of inputs. It thus reduces the number of transmission lines and can be compared to a multiplexer only one of the inputs become "high" (logic state "1") at a time.

Some typical examples would be 4:2 encoder, 8:3 encoder, 16:4 encoders etc.

4-to-2 Encoder:

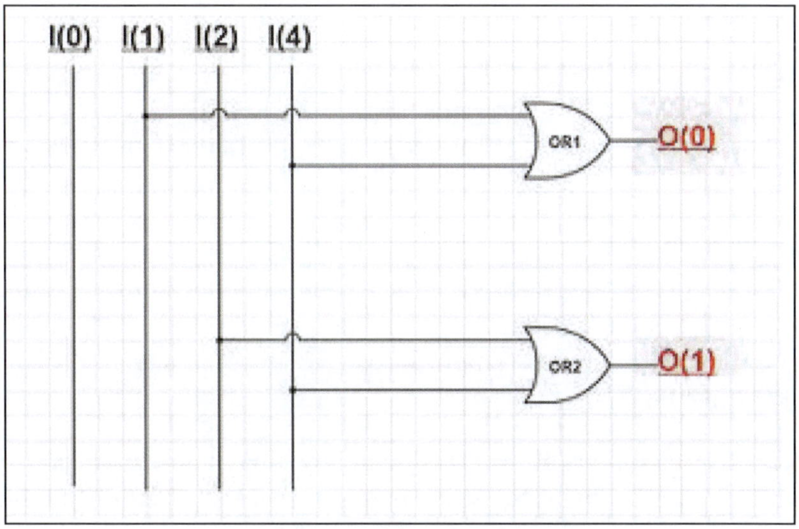

A simple 4:2 Encoder using OR gate.

8-to-3 Encoder:

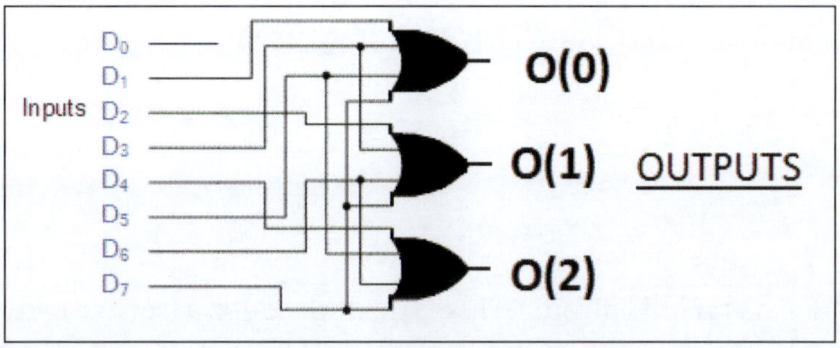

The image represents a 8:3 encoder. 4-input OR gates have been used for simplicity; one can also use 2 2-input OR gates.

A Truth Table representing 8:3 Encoder.										
INPUT								OUTPUT		
a	a	a	a	a	a	a	a	d	d	d
0	0	0	0	0	0	0	0	X	X	X
0	0	0	0	0	0	0	1	0	0	0
0	0	0	0	0	0	1	0	0	0	1
0	0	0	0	0	1	0	0	0	1	0
0	0	0	0	1	0	0	0	0	1	1
0	0	0	1	0	0	0	0	1	0	0
0	0	1	0	0	0	0	0	1	0	1
0	1	0	0	0	0	0	0	1	1	0
1	0	0	0	0	0	0	0	1	1	1

Binary Decoder

In digital electronics, a binary decoder is a combinational logic circuit that converts binary information from the n coded inputs to a maximum of 2^n unique outputs. They are used in a wide variety of applications, including data demultiplexing, seven segment displays, and memory address decoding.

There are several types of binary decoders, but in all cases a decoder is an electronic circuit with multiple input and multiple output signals, which converts every unique combination of input states to a specific combination of output states. In addition to integer data inputs, some decoders also have one or more "enable" inputs. When the enable input is negated (disabled), all decoder outputs are forced to their inactive states.

Depending on its function, a binary decoder will convert binary information from n input signals to as many as 2^n unique output signals. Some decoders have less than 2^n output lines; in such cases, at least one output pattern may be repeated for different input values.

A binary decoder is usually implemented as either a stand-alone integrated circuit (IC) or as part of a more complex IC. In the latter case the decoder may be synthesized by means of a hardware description language such as VHDL or Verilog. Widely used decoders are often available in the form of standardized ICs.

Types of Decoders

1-of-n Decoder

A 1-of-n binary decoder has n output bits. This type of decoder asserts exactly one of its n output bits, or none of them, for every integer input value. The "address" (bit number) of the activated output is specified by the integer input value. For example, output bit number 0 is selected when the integer value 0 is applied to the inputs.

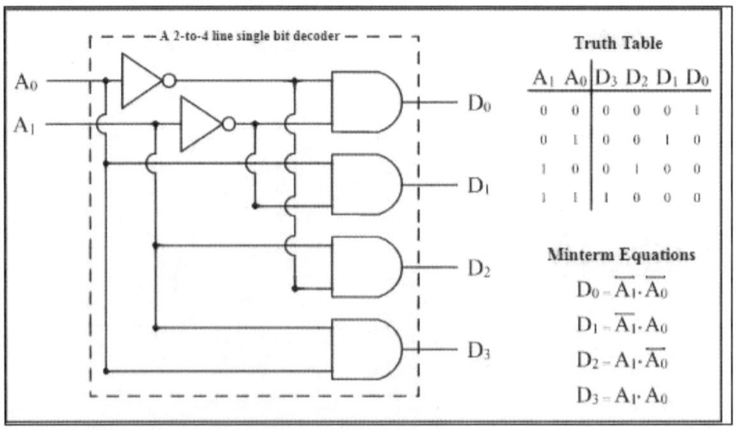

A 2-to-4 line decoder.

Examples of this type of decoder include:

- A *3-to-8 line decoder* activates one of eight output bits for each input value from 0 to 7

— the range of integer values that can be expressed in three bits. Similarly, a *4-to-16 line decoder* activates one of 16 outputs for each 4-bit input in the integer range [0,15].

- A *BCD to decimal decoder* has ten output bits. It accepts an input value consisting of a binary-coded decimal integer value and activates one specific, unique output for every input value in the range [0,9]. All outputs are held inactive when a non-decimal value is applied to the inputs.
- A demultiplexer is a 1-of-n binary decoder that is used to route a data bit to one of its n outputs while all other outputs remain inactive.

Code Translator

Code translators differ from 1-of-n decoders in that multiple output bits may be active at the same time. An example of this is a *seven-segment decoder*, which converts an integer into the combination of segment control signals needed to display the integer's value on a seven-segment display digit.

One variant of seven-segment decoder is the *BCD to seven-segment decoder*, which translates a binary-coded decimal value into the corresponding segment control signals for input integer values 0 to 9. This decoder function is available in standard ICs such as the CMOS 4511.

Programmable Logic Device

A programmable logic device (PLD) is an electronic component used to build reconfigurable digital circuits. Unlike integrated circuits (IC) which consist of logic gates and have a fixed function, a PLD has an undefined function at the time of manufacture. Before the PLD can be used in a circuit it must be programmed (reconfigured) by using a specialized program.

In 1969, Motorola offered the XC157, a mask-programmed gate array with 12 gates and 30 uncommitted input/output pins.

In 1970, Texas Instruments developed a mask-programmable IC based on the IBM read-only associative memory or ROAM. This device, the TMS2000, was programmed by altering the metal layer during the production of the IC. The TMS2000 had up to 17 inputs and 18 outputs with 8 JK flip flop for memory. TI coined the term Programmable Logic Array (PLA) for this device.

In 1971, General Electric Company (GE) was developing a programmable logic device based on the new Programmable Read-Only Memory (PROM) technology. This experimental device improved on IBM's ROAM by allowing multilevel logic. Intel had just introduced the floating-gate UV erasable PROM so the researcher at GE incorporated that technology. The GE device was the first erasable PLD ever developed, predating the Altera EPLD by over a decade. GE obtained several early patents on programmable logic devices.

In 1973 National Semiconductor introduced a mask-programmable PLA device (DM7575) with 14 inputs and 8 outputs with no memory registers. This was more popular than the TI part but cost of making the metal mask limited its use. The device is significant because it was the basis for the field programmable logic array produced by Signetics in 1975, the 82S100. (Intersil actually beat Signetics to market but poor yield doomed their part.)

In 1974 GE entered into an agreement with Monolithic Memories (MMI) to develop a mask–programmable logic device incorporating the GE innovations. The device was named the 'Programmable Associative Logic Array' or PALA. The MMI 5760 was completed in 1976 and could implement multilevel or sequential circuits of over 100 gates. The device was supported by a GE design environment where Boolean equations would be converted to mask patterns for configuring the device. The part was never brought to market.

PLA

In 1970, Texas Instruments developed a mask-programmable IC based on the IBM read-only associative memory or ROAM. This device, the TMS2000, was programmed by altering the metal layer during the production of the IC. The TMS2000 had up to 17 inputs and 18 outputs with 8 JK flip-flops for memory. TI coined the term programmable logic array for this device.

A programmable logic array (PLA) has a programmable AND gate array, which links to a programmable OR gate array, which can then be conditionally complemented to produce an output.

PAL

PAL devices have arrays of transistor cells arranged in a "fixed-OR, programmable-AND" plane used to implement "sum-of-products" binary logic equations for each of the outputs in terms of the inputs and either synchronous or asynchronous feedback from the outputs.

MMI introduced a breakthrough device in 1978, the programmable array logic or PAL. The architecture was simpler than that of Signetics FPLA because it omitted the programmable OR array. This made the parts faster, smaller and cheaper. They were available in 20 pin 300 mil DIP packages while the FPLAs came in 28 pin 600 mil packages. The PAL Handbook demystified the design process. The PALASM design software (PAL assembler) converted the engineers' Boolean equations into the fuse pattern required to program the part. The PAL devices were soon second-sourced by National Semiconductor, Texas Instruments and AMD.

After MMI succeeded with the 20-pin PAL parts, AMD introduced the 24-pin 22V10 PAL with additional features. After buying out MMI (1987), AMD spun off a consolidated operation as Vantis, and that business was acquired by Lattice Semiconductor in 1999.

GALs

An improvement on the PAL was the generic array logic device, or GAL, invented by Lattice Semiconductor in 1985. This device has the same logical properties as the PAL but can be erased and reprogrammed. The GAL is very useful in the prototyping stage of a design, when any bugs in the logic can be corrected by reprogramming. GALs are programmed and reprogrammed using a PAL programmer, or by using the in-circuit programming technique on supporting chips.

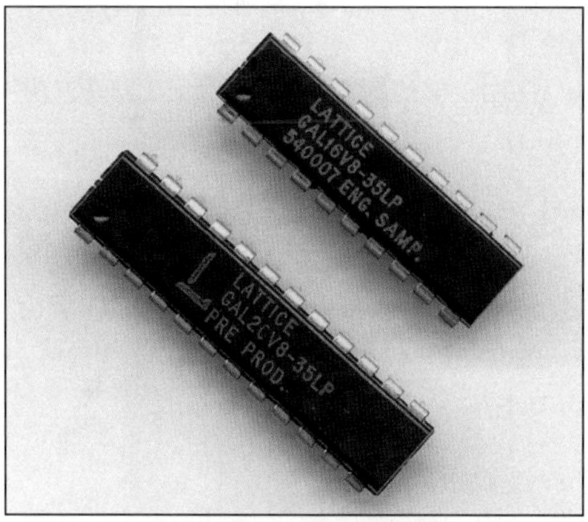

Lattice GAL 16V8 and 20V8.

Lattice GALs combine CMOS and electrically erasable (E^2) floating gate technology for a high-speed, low-power logic device. A similar device called a PEEL (programmable electrically erasable logic) was introduced by the International CMOS Technology (ICT) corporation.

CPLDs

PALs and GALs are available only in small sizes, equivalent to a few hundred logic gates. For bigger logic circuits, complex PLDs or CPLDs can be used. These contain the equivalent of several PALs linked by programmable interconnections, all in one integrated circuit. CPLDs can replace thousands, or even hundreds of thousands, of logic gates.

Some CPLDs are programmed using a PAL programmer, but this method becomes inconvenient for devices with hundreds of pins. A second method of programming is to solder the device to its printed circuit board, then feed it with a serial data stream from a personal computer. The CPLD contains a circuit that decodes the data stream and configures the CPLD to perform its specified logic function. Some manufacturers (including Altera and Microsemi) use JTAG to program CPLDs in-circuit from .JAM files.

FPGAs

While PALs were being developed into GALs and CPLDs (all discussed), a separate stream of development was happening. This type of device is based on gate array technology and is called the field-programmable gate array (FPGA). Early examples of FPGAs are the 82s100 array, and 82S105 sequencer, by Signetics, introduced in the late 1970s. The 82S100 was an array of AND terms. The 82S105 also had flip flop functions.

(Anmerk: 82S100 und ähnl ICs from Signetics have PLA-Structure, AND-Plane + OR-Plane).

FPGAs use a grid of logic gates, and once stored, the data doesn't change, similar to that of an ordinary gate array. The term "field-programmable" means the device is programmed by the customer, not the manufacturer.

FPGAs are usually programmed after being soldered down to the circuit board, in a manner similar to that of larger CPLDs. In most larger FPGAs, the configuration is volatile and must be re-loaded into the device whenever power is applied or different functionality is required. Configuration is typically stored in a configuration PROM or EEPROM. EEPROM versions may be in-system programmable (typically via JTAG).

The difference between FPGAs and CPLDs is that FPGAs are internally based on look-up tables (LUTs) whereas CPLDs form the logic functions with sea-of-gates (e.g. sum of products). CPLDs are meant for simpler designs while FPGAs are meant for more complex designs. In general, CPLDs are a good choice for wide combinational logic applications, whereas FPGAs are more suitable for large state machines such as microprocessors.

Other Variants

These are microprocessor circuits that contain some fixed functions and other functions that can be altered by code running on the processor. Designing self-altering systems requires engineers to learn new methods, and that new software tools be developed.

PLDs are being sold now that contain a microprocessor with a fixed function (the so-called *core*) surrounded by programmable logic. These devices let designers concentrate on adding new features to designs without having to worry about making the microprocessor work. Also, the fixed-function microprocessor takes less space on the chip than a part of the programmable gate array implementing the same processor, leaving more space for the programmable gate array to contain the designer's specialized circuits.

How PLDs Retain their Configuration

A PLD is a combination of a logic device and a memory device. The memory is used to store the pattern that was given to the chip during programming. Most of the methods for storing data in an integrated circuit have been adapted for use in PLDs. These include:

- Silicon antifuses;
- SRAM;
- EPROM or EEPROM cells;
- Flash memory.

Silicon antifuses are connections that are made by applying a voltage across a modified area of silicon inside the chip. They are called antifuses because they work in the opposite way to normal fuses, which begin life as connections until they are broken by an electric current.

SRAM, or static RAM, is a volatile type of memory, meaning that its contents are lost each time the power is switched off. SRAM-based PLDs therefore have to be programmed every time the circuit is switched on. This is usually done automatically by another part of the circuit.

An EPROM cell is a MOS (metal-oxide-semiconductor) transistor that can be switched on by trapping an electric charge permanently on its gate electrode. This is done by a PAL programmer. The charge remains for many years and can only be removed by exposing the chip to strong ultraviolet light in a device called an EPROM eraser.

Flash memory is non-volatile, retaining its contents even when the power is switched off. It can be erased and reprogrammed as required. This makes it useful in PLDs that may be reprogrammed frequently, such as PLDs used in prototypes. Flash memory is a kind of EEPROM that holds information using trapped electric charges similar to EPROM. Consequently, Flash memory can hold information for years, but possibly not as many years as EPROM.

As of 2005, most CPLDs are electrically programmable and erasable, and non-volatile. This is because they are too small to justify the inconvenience of programming internal SRAM cells every time they start up, and EPROM cells are more expensive due to their ceramic package with a quartz window.

PLD Programming Languages

Many PAL programming devices accept input in a standard file format, commonly referred to as 'JEDEC files'. They are analogous to software compilers. The languages used as source code for logic compilers are called hardware description languages, or HDLs.

PALASM, ABEL and CUPL are frequently used for low-complexity devices, while Verilog and VHDL are popular higher-level description languages for more complex devices. The more limited ABEL is often used for historical reasons, but for new designs VHDL is more popular, even for low-complexity designs.

PLD Programming Devices

A device programmer is used to transfer the boolean logic pattern into the programmable device. In the early days of programmable logic, every PLD manufacturer also produced a specialized device programmer for its family of logic devices. Later, universal device programmers came onto the market that supported several logic device families from different manufacturers. Today's device programmers usually can program common PLDs (mostly PAL/GAL equivalents) from all existing manufacturers. Common file formats used to store the boolean logic pattern (fuses) are JEDEC, Altera POF (programmable object file), or Xilinx BITstream.

Seven-segment Display

A seven-segment display (SSD), or seven-segment indicator, is a form of electronic display device for displaying decimal numerals that is an alternative to the more complex dot matrix displays.

Seven-segment displays are widely used in digital clocks, electronic meters, basic calculators, and other electronic devices that display numerical information.

Concept and Visual Structure

The common segment displays shown side by side: 7-segment, 9-segment, 14-segment and 16-segment displays.

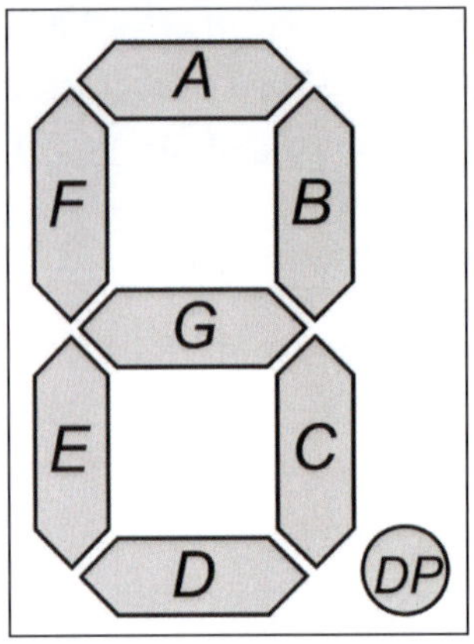

The individual segments of a seven-segment display.

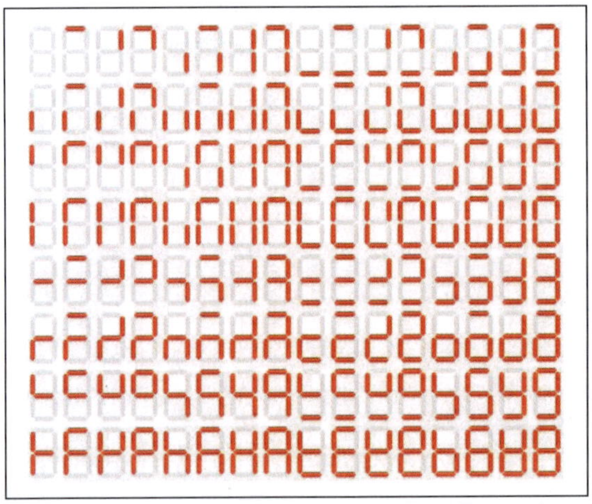

16x8-grid showing the 128 states of a seven-segment display.

The seven elements of the display can be lit in different combinations to represent the Arabic numerals. Often the seven segments are arranged in an *oblique* (slanted) arrangement, which aids readability. In most applications, the seven segments are of nearly uniform shape and size (usually elongated hexagons, though trapezoids and rectangles can also be used), though in the case of adding machines, the vertical segments are longer and more oddly shaped at the ends in an effort to further enhance readability.

The numerals 6 and 9 may be represented by two different glyphs on seven-segment displays, with or without a 'tail'. The numeral 7 also has two versions, with or without segment F.

The seven segments are arranged as a rectangle of two vertical segments on each side with one horizontal segment on the top, middle, and bottom. Additionally, the seventh segment bisects the rectangle horizontally. There are also fourteen-segment displays and sixteen-segment displays (for full alphanumerics); however, these have mostly been replaced by dot matrix displays. Twenty-two segment displays capable of displaying the full ASCII character set were briefly available in the early 1980s, but did not prove popular.

The segments of a 7-segment display are referred to by the letters A to G, where the optional decimal point (an "eighth segment", referred to as DP) is used for the display of non-integer numbers.

Implementations

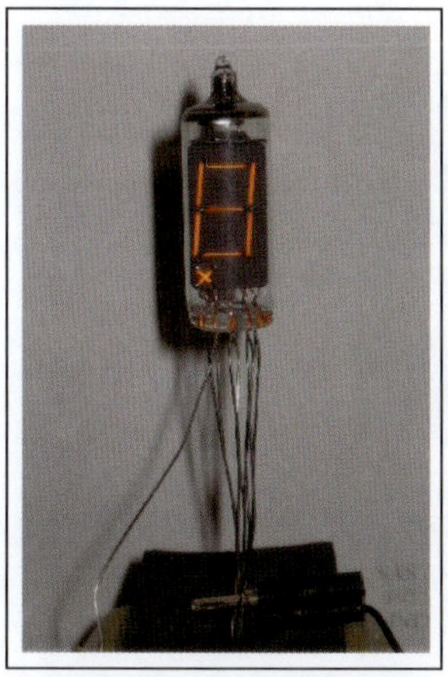		
An incandescent filament-type early seven-segment display, a.k.a. *Numitron*.	A mechanical seven-segment display for displaying automotive fuel prices.	Typical pinout of a common cathode electronic seven-segment display.

Seven-segment displays may use a liquid crystal display (LCD), a light-emitting diode (LED) for each segment, an electrochromic display, or other light-generating or controlling techniques such as cold cathode gas discharge (Panaplex), vacuum fluorescent, incandescent filaments (Numitron), and others. For gasoline price totems and other large signs, vane displays made up of electromagnetically flipped light-reflecting segments (or "vanes") are still commonly used. An alternative to the 7-segment display in the 1950s through the 1970s was the cold-cathode, neon-lamp-like nixie tube. Starting in 1970, RCA sold a display device known as the *Numitron* that used incandescent filaments arranged into a seven-segment display.

In a simple LED package, typically all of the cathodes (negative terminals) or all of the anodes (positive terminals) of the segment LEDs are connected and brought out to a common pin; this is referred to as a "common cathode" or "common anode" device. Hence a 7 segment plus decimal point package will only require nine pins, though commercial products typically contain more pins, and/or spaces where pins would go, in order to match standard IC sockets. Integrated displays also exist, with single or multiple digits. Some of these integrated displays incorporate their own internal decoder, though most do not: each individual LED is brought out to a connecting pin as described.

Multiple-digit LED displays as used in pocket calculators and similar devices used multiplexed displays to reduce the number of I/O pins required to control the display. For example, all the anodes of the A segments of each digit position would be connected together and to a driver circuit pin, while the cathodes of all segments for each digit would be connected. To operate any particular segment of any digit, the controlling integrated circuit would turn on the cathode driver for the selected digit, and the anode drivers for the desired segments; then after a short blanking interval the next digit would be selected and new segments lit, in a sequential fashion. In this manner an eight digit display with seven segments and a decimal point would require only 8 cathode drivers and 8 anode drivers, instead of sixty-four drivers and IC pins. Often in pocket calculators the digit drive lines would be used to scan the keyboard as well, providing further savings; however, pressing multiple keys at once would produce odd results on the multiplexed display.

A multiplexed 4-digit, seven-segment display with only 12 pins.

Although to a naked eye all digits of an LED display appear lit, the implementation of a typical multiplexed display described above means that in reality only a single digit is lit at any given time.

A single byte can encode the full state of a 7-segment-display. The most popular bit encodings are *gfedcba* and *abcdefg*, where each letter represents a particular segment in the display. In the *gfedcba* representation, a byte value of 0x06 would (in a common anode circuit) turn on segments 'c' and 'b', which would display a '1'.

Displaying Letters

Hexadecimal digits can be displayed on seven-segment displays. Today, a combination of uppercase and lowercase letters is commonly used for A–F; this is done to obtain a unique, unambiguous shape for each hexadecimal digit (otherwise, a capital 'D' would look identical to a '0' and a capital 'B' would look identical to an '8'). Also the digit '6' must be displayed with the top bar lit to avoid ambiguity with the letter 'b'.

Hexadecimal encodings for displaying the digits 0 to F										
Digit	Display	gfedcba	abcdefg	a	b	c	d	e	f	g
0		0x3F	0x7E	on	on	on	on	on	on	off
1		0x06	0x30	off	on	on	off	off	off	off
2		0x5B	0x6D	on	on	off	on	on	off	on
3		0x4F	0x79	on	on	on	on	off	off	on
4		0x66	0x33	off	on	on	off	off	on	on
5		0x6D	0x5B	on	off	on	on	off	on	on
6		0x7D	0x5F	on	off	on	on	on	on	on
7		0x07	0x70	on	on	on	off	off	off	off
8		0x7F	0x7F	on	on	on	on	on	on	on
9		0x6F	0x7B	on	on	on	on	off	on	on
A		0x77	0x77	on	on	on	off	on	on	on

b	b	0x7C	0x1F	off	off	on	on	on	on	on
C	C	0x39	0x4E	on	off	off	on	on	on	off
d	d	0x5E	0x3D	off	on	on	on	on	off	on
E	E	0x79	0x4F	on	off	off	on	on	on	on
F	F	0x71	0x47	on	off	off	off	on	on	on

However, this modern scheme wasn't always followed in the past, and various other schemes could be found as well:

- The Texas Instruments seven-segment display decoder chips 7446/7447/7448/7449 and 74246/74247/74248/74249 and the Siemens FLH551-7448/555-8448 chips used truncated versions of "2", "3", "4", "5" and "6" for digits A–E. Digit F (1111 binary) was blank.
- Soviet programmable calculators like the Б3-34 instead used the symbols "–", "L", "C", "Г", "E", and " " (space) to display hexadecimal numbers above nine, allowing the error message EГГoГ to be displayed.
- Not all 7-segment decoders were suitable to display digits above nine at all. Many earlier chips provided logic designed for only for 0–9 and higher numbers produced whatever pattern resulted. For comparison, the National Semiconductor MM74C912 displayed "o" for A and B, "–" for C, D and E, and blank for F. The CD4511 even just displayed blanks.

In addition, seven-segment displays can be used to show various other letters of the Latin, Cyrillic and Greek alphabets including punctuation, but few representations are unambiguous and intuitive at the same time. Short messages giving status information (e.g. "no dISC" on a CD player) are also commonly represented on 7-segment displays. In the case of such messages it is not necessary for every letter to be unambiguous, merely for the words as a whole to be readable.

Similar displays with fourteen or sixteen segments are available allowing less-ambiguous representations of the alphabet.

Using a restricted range of letters that look like (upside-down) digits, seven-segment displays are commonly used by school children to form words and phrases using a technique known as "calculator spelling".

Sequential Logic

In digital circuit theory, sequential logic is a type of logic circuit whose output depends not only on the present value of its input signals but on the sequence of past inputs, the input history as well. This is in contrast to *combinational logic*, whose output is a function of only the present input. That is, sequential logic has *state* (*memory*) while combinational logic does not.

Sequential logic is used to construct finite state machines, a basic building block in all digital circuitry. Virtually all circuits in practical digital devices are a mixture of combinational and sequential logic.

A familiar example of a device with sequential logic is a television set with "channel up" and "channel down" buttons. Pressing the "up" button gives the television an input telling it to switch to the next channel above the one it is currently receiving. If the television is on channel 5, pressing "up" switches it to receive channel 6. However, if the television is on channel 8, pressing "up" switches it to channel "9". In order for the channel selection to operate correctly, the television must be aware of which channel it is currently receiving, which was determined by past channel selections. The television stores the current channel as part of its *state*. When a "channel up" or "channel down" input is given to it, the sequential logic of the channel selection circuitry calculates the new channel from the input and the current channel.

Digital sequential logic circuits are divided into synchronous and asynchronous types. In synchronous sequential circuits, the state of the device changes only at discrete times in response to a clock signal. In asynchronous circuits the state of the device can change at any time in response to changing inputs.

Synchronous Sequential Logic

Nearly all sequential logic today is *clocked* or *synchronous* logic. In a synchronous circuit, an electronic oscillator called a *clock* (or clock generator) generates a sequence of repetitive pulses called the *clock signal* which is distributed to all the memory elements in the circuit. The basic memory element in sequential logic is the flip-flop. The output of each flip-flop only changes when triggered by the clock pulse, so changes to the logic signals throughout the circuit all begin at the same time, at regular intervals, synchronized by the clock.

The output of all the storage elements (flip-flops) in the circuit at any given time, the binary data they contain, is called the *state* of the circuit. The state of a synchronous circuit only changes on clock pulses. At each cycle, the next state is determined by the current state and the value of the input signals when the clock pulse occurs.

The main advantage of synchronous logic is its simplicity. The logic gates which perform the operations on the data require a finite amount of time to respond to changes to their inputs. This is called *propagation delay*. The interval between clock pulses must be long enough so that all the logic gates have time to respond to the changes and their outputs "settle" to stable logic values, before the next clock pulse occurs. As long as this condition is met (ignoring certain other details) the circuit is guaranteed to be stable and reliable. This determines the maximum operating speed of a synchronous circuit.

Synchronous logic has two main disadvantages:

- The maximum possible clock rate is determined by the slowest logic path in the circuit, otherwise known as the critical path. Every logical calculation, from the simplest to the most complex, must complete in one clock cycle. So logic paths that complete their calculations quickly are idle much of the time, waiting for the next clock pulse. Therefore, synchronous logic can be slower than asynchronous logic. One way to speed up synchronous circuits is to split complex operations into several simple operations which can be performed in successive clock cycles, a technique known as *pipelining*. This technique is extensively used in microprocessor design, and helps to improve the performance of modern processors.

- The clock signal must be distributed to every flip-flop in the circuit. As the clock is usually a high-frequency signal, this distribution consumes a relatively large amount of power and dissipates much heat. Even the flip-flops that are doing nothing consume a small amount of power, thereby generating waste heat in the chip. In portable devices which have limited battery power, the clock signal goes on even when the device is not being used, consuming power.

Asynchronous Sequential Logic

Asynchronous sequential logic is not synchronized by a clock signal; the outputs of the circuit change directly in response to changes in inputs. The advantage of asynchronous logic is that it can be faster than synchronous logic, because the circuit doesn't have to wait for a clock signal to process inputs. The speed of the device is potentially limited only by the propagation delays of the logic gates used.

However, asynchronous logic is more difficult to design and is subject to problems not encountered in synchronous designs. The main problem is that digital memory elements are sensitive to the order that their input signals arrive; if two signals arrive at a flip-flop or latch at almost the same time, which state the circuit goes into can depend on which signal gets to the gate first. Therefore, the circuit can go into the wrong state, depending on small differences in the propagation delays of the logic gates. This is called a race condition. This problem is not as severe in synchronous circuits because the outputs of the memory elements only change at each clock pulse. The interval between clock signals is designed to be long enough to allow the outputs of the memory elements to "settle" so they are not changing when the next clock comes. Therefore, the only timing problems are due to "asynchronous inputs"; inputs to the circuit from other systems which are not synchronized to the clock signal.

Asynchronous sequential circuits are typically used only in a few critical parts of otherwise synchronous systems where speed is at a premium, such as parts of microprocessors and digital signal processing circuits.

The design of asynchronous logic uses different mathematical models and techniques from synchronous logic, and is an active area of research.

Latch

A latch is an electronic logic circuit that has two inputs and one output. One of the inputs is called the SET input; the other is called the RESET input.

Latch circuits can be either active-high or active-low. The difference is determined by whether the operation of the latch circuit is triggered by HIGH or LOW signals on the inputs.

- Active-high circuit: Both inputs are normally tied to ground (LOW), and the latch is triggered by a momentary HIGH signal on either of the inputs.

- Active-low circuit: Both inputs are normally HIGH, and the latch is triggered by a momentary LOW signal on either input.

In an active-high latch, both the SET and RESET inputs are connected to ground. When the SET input goes HIGH, the output also goes HIGH. When the SET input returns to LOW, however, the output remains HIGH. The output of the active-high latch stays HIGH until the RESET input goes HIGH. Then, the output returns to LOW and will go HIGH again only when the SET input is triggered once more.

In other words, the latch remembers that the SET input has been activated. If the SET input goes HIGH for even a moment, the output goes HIGH and stays HIGH, even after the SET input returns to LOW. The output returns to LOW only when the RESET input goes HIGH.

On the other hand, in an active-low latch the inputs are normally held at HIGH. When the SET input momentarily goes LOW, the output goes HIGH. The output then stays HIGH until the RESET input momentarily goes LOW.

Note that most latch circuits actually have a second output that is simply the first output inverted. In other words, whenever the first output is HIGH, the second output is LOW, and vice versa. These outputs are usually referred to as Q and Q-bar with the latter notated as follows:

$$\overline{Q}$$

The horizontal bar symbol over a label is a common logical shorthand for inversion. That is, Q-bar is the inverse of Q. If Q is HIGH, Q-bar is LOW, and if Q is LOW, Q-bar is HIGH.

You can easily create an active-high latch from a pair of NOR gates. (The output of a NOR gate is HIGH if both inputs are LOW; otherwise, the output is LOW.) In this circuit, the SET input is connected to one of the inputs of the first NOR gate, and the RESET input is connected to one of the inputs of the second NOR gate.

The trick of the latch circuit is that the output of the NOR gates are cross-connected to the remaining NOR gate inputs. In other words, the output from the first NOR gate is connected to one of the inputs of the second NOR gate, and the output from the second NOR gate is connected to one of the inputs of the first NOR gate.

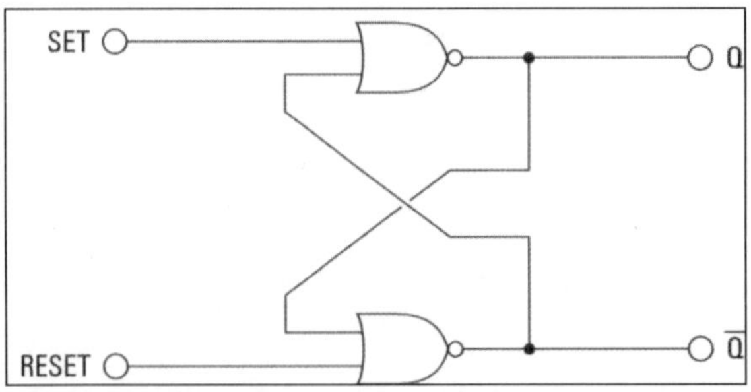

The next schematic is for an active-low latch. The only difference between this schematic and the one shown previously is that the active-low latch uses NAND gates instead of NOR gates. Notice also in this diagram that the inputs are referred to as SET-bar and RESET-bar rather than SET and RESET, which indicates that the inputs are active-low.

Flip Flops

In electronics, a flip-flop or latch is a circuit that has two stable states and can be used to store state information – a bistable multivibrator. The circuit can be made to change state by signals applied to one or more control inputs and will have one or two outputs. It is the basic storage element in sequential logic. Flip-flops and latches are fundamental building blocks of digital electronics systems used in computers, communications, and many other types of systems.

Flip-flops and latches are used as data storage elements. A flip-flop is a device which stores a single *bit* (binary digit) of data; one of its two states represents a "one" and the other represents a "zero". Such data storage can be used for storage of *state*, and such a circuit is described as sequential logic in electronics. When used in a finite-state machine, the output and next state depend not only on its current input, but also on its current state (and hence, previous inputs). It can also be used for counting of pulses, and for synchronizing variably-timed input signals to some reference timing signal.

Flip-flops can be either level-triggered (asynchronous, transparent or opaque) or edge-triggered (synchronous, or clocked). The term flip-flop has historically referred generically to both level-triggered and edge-triggered circuits that store a single bit of data using gates. Recently, some authors reserve the term *flip-flop* exclusively for discussing clocked circuits; the simple ones are commonly called *transparent latches*,. Using this terminology, a level-sensitive flip-flop is called a transparent latch, whereas an edge-triggered flip-flop is simply called a flip-flop. Using either terminology, the term "flip-flop" refers to a device that stores a single bit of data, but the term "latch" may also refer to a device that stores any number of bits of data using a single trigger. The terms "edge-triggered", and "level-triggered" may be used to avoid ambiguity.

When a level-triggered latch is enabled it becomes transparent, but an edge-triggered flip-flop's output only changes on a single type (positive going or negative going) of clock edge.

Implementation

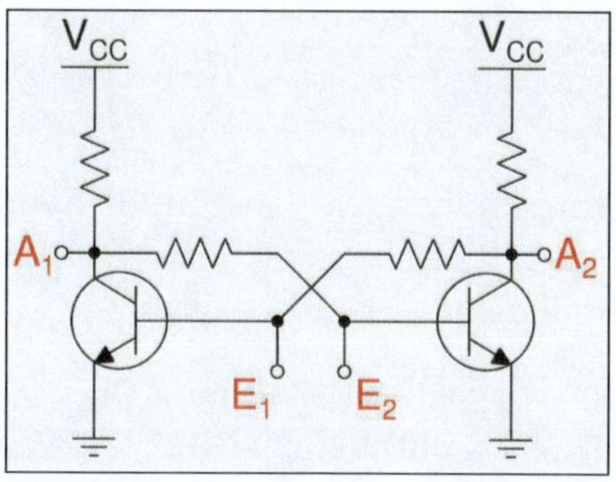

A traditional (simple) flip-flop circuit based on bipolar junction transistors.

Flip-flops can be either simple (transparent or asynchronous) or clocked (synchronous). In the context of hardware description languages, the simple ones are commonly described as *latches*, while the clocked ones are described as *flip-flops*.

Simple flip-flops can be built around a single pair of cross-coupled inverting elements: vacuum tubes, bipolar transistors, field effect transistors, inverters, and inverting logic gates have all been used in practical circuits.

Clocked devices are specially designed for synchronous systems; such devices ignore their inputs except at the transition of a dedicated clock signal (known as clocking, pulsing, or strobing). Clocking causes the flip-flop either to change or to retain its output signal based upon the values of the input signals at the transition. Some flip-flops change output on the rising edge of the clock, others on the falling edge.

Since the elementary amplifying stages are inverting, two stages can be connected in succession (as a cascade) to form the needed non-inverting amplifier. In this configuration, each amplifier may be considered as an active inverting feedback network for the other inverting amplifier. Thus the two stages are connected in a non-inverting loop although the circuit diagram is usually drawn as a symmetric cross-coupled pair (both the drawings are initially introduced in the Eccles–Jordan patent).

Flip-flop Types

Flip-flops can be divided into common types: the SR ("set-reset"), D ("data" or "delay"), T ("toggle"), and JK. The behavior of a particular type can be described by what is termed the characteristic equation, which derives the "next" (i.e., after the next clock pulse) output, Q_{next} in terms of the input signal(s) and/or the current output.

Simple Set-reset Latches

SR NOR Latch

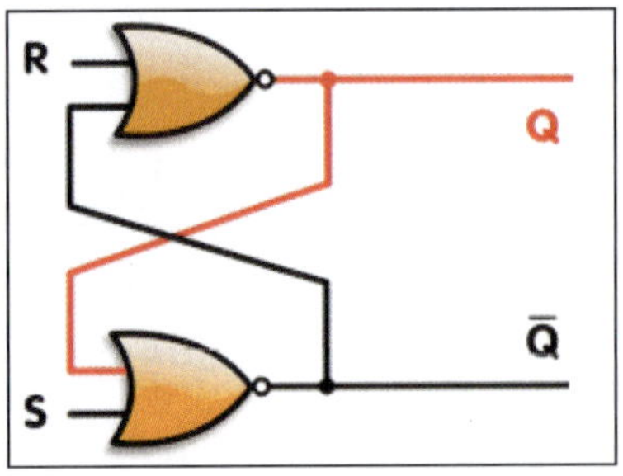

SR latch, constructed from a pair of cross-coupled NOR gates.
Red and black mean logical '1' and '0', respectively.

When using static gates as building blocks, the most fundamental latch is the simple *SR latch*, where S and R stand for *set* and *reset*. It can be constructed from a pair of cross-coupled NOR logic gates. The stored bit is present on the output marked Q.

While the R and S inputs are both low, feedback maintains the Q and $\overline{Q}$ outputs in a constant state, with $\overline{Q}$ the complement of Q. If S (*Set*) is pulsed high while R (*Reset*) is held low, then the Q output is forced high, and stays high when S returns to low; similarly, if R is pulsed high while S is held low, then the Q output is forced low, and stays low when R returns to low.

SR latch operation							
Characteristic table				Excitation table			
S	R	Q_{next}	Action	Q	Q_{next}	S	R
0	0	Q	hold state	0	0	0	X

0	1	0	reset	0	1	1	0
1	0	1	set	1	0	0	1
1	1	X	not allowed	1	1	X	0

The R = S = 1 combination is called a restricted combination or a forbidden state because, as both NOR gates then output zeros, it breaks the logical equation Q = not $\overline{Q}$. The combination is also inappropriate in circuits where *both* inputs may go low *simultaneously* (i.e. a transition from *restricted* to *keep*). The output would lock at either 1 or 0 depending on the propagation time relations between the gates (a race condition).

To overcome the restricted combination, one can add gates to the inputs that would convert (S,R) = (1,1) to one of the non-restricted combinations. That can be:

- Q = 1 (1,0) – referred to as an *S (dominated)-latch;*
- Q = 0 (0,1) – referred to as an *R (dominated)-latch;*

  This is done in nearly every programmable logic controller.
- Keep state (0,0) – referred to as an *E-latch.*

Alternatively, the restricted combination can be made to *toggle* the output. The result is the JK latch.

The characteristic equation for the SR latch is :

$$Q_{\text{next}} = \bar{R}Q + \bar{R}S \text{ or } Q_{\text{next}} = \bar{R}(Q + S).$$

Another expression is:

$$Q_{\text{next}} = S + \bar{R}Q \text{ with } SR = 0.$$

$\overline{\text{SR}}$ NAND Latch

When using static gates as building blocks, the most fundamental latch is the simple SR latch, where S and R stand for set and reset. It can be constructed from a pair of cross-coupled NOR or NAND logic gates. The stored bit is present on the output marked Q.

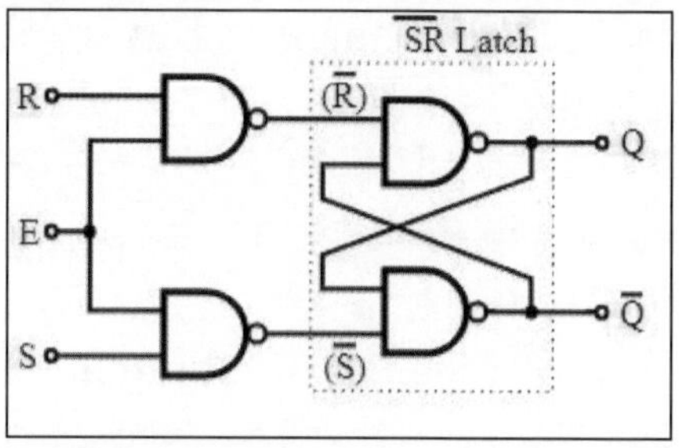

An SR latch constructed from cross-coupled NAND gates.

The circuit shown below is a basic NAND latch. The inputs are generally designated S and R for Set and Reset respectively. Because the NAND inputs must normally be logic 1 to avoid affecting the latching action, the inputs are considered to be inverted in this circuit (or active low).

The circuit uses feedback to "remember" and retain its logical state even after the controlling input signals have changed. When the S and R inputs are both high, feedback maintains the Q outputs to the previous state.

$\overline{SR}$ latch operation		
$\overline{S}$	$\overline{R}$	Action
0	0	Q=1, not(Q)=1, Not allowed
0	1	Q = 1
1	0	Q = 0
1	1	No change & random start

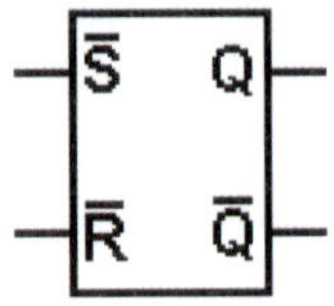

Symbol for an $\overline{SR}$ NAND latch.

SR AND-OR Latch

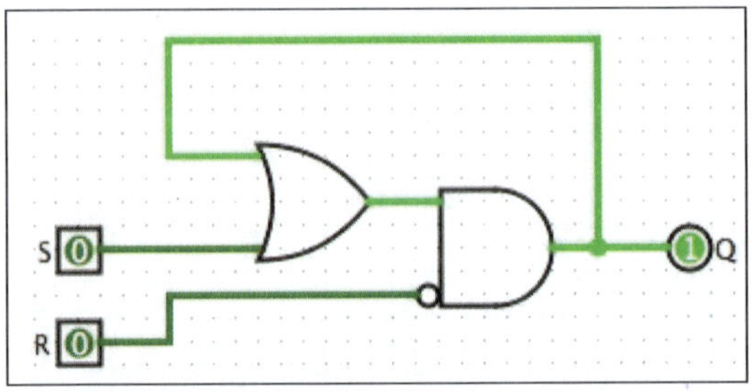

An SR AND-OR latch. Light green means logical '1' and dark green means logical '0'. The latch is currently in hold mode (no change).

SR latches realised as a pair of cross-coupled components (transistors, gates, tubes, etc.) are rather hard to understand for beginners. A didactically easier to understand model uses a single feedback loop instead of the cross-coupling. The following is an SR latch built with an AND gate with one inverted input and an OR gate.

SR AND-OR latch operation		
S	R	Action
0	0	No change & random start
1	0	Q = 1
X	1	Q = 0

Note that the SR AND-OR latch has the benefit that S=1,R=1 is well defined. In above version of the SR AND-OR latch it gives priority to the R signal over the S signal. If priority of S over R is needed, this can be achieved by changing the order of the AND and OR gate.

JK Latch

The JK latch is much less frequently used than the JK flip-flop. The JK latch follows the following state table:

JK latch truth table			
J	K	Q_{next}	Comment
0	0	Q	No change

0	1	0	Reset
1	0	1	Set
1	1	$\overline{Q}$	Toggle

Hence, the JK latch is an SR latch that is made to *toggle* its output (oscillate between 0 and 1) when passed the input combination of 11. Unlike the JK flip-flop, the 11 input combination for the JK latch is not very useful because there is no clock that directs toggling.

Gated Latches and Conditional Transparency

Latches are designed to be *transparent*. That is, input signal changes cause immediate changes in output. Additional logic can be added to a simple transparent latch to make it *non-transparent* or *opaque* when another input (an "enable" input) is not asserted. When several *transparent* latches follow each other, using the same enable signal, signals can propagate through all of them at once. However, by following a *transparent-high* latch with a *transparent-low* (or *opaque-high*) latch, a master–slave flip-flop is implemented.

Gated SR Latch

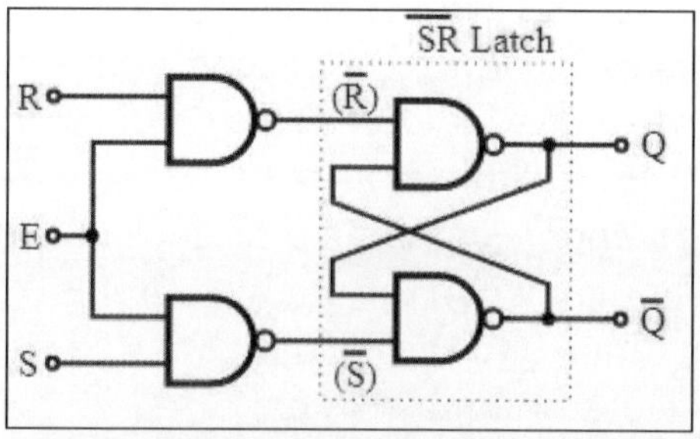

Gated SR latch constructed from four NAND gates with the original $\overline{SR}$ NAND latch encircled.

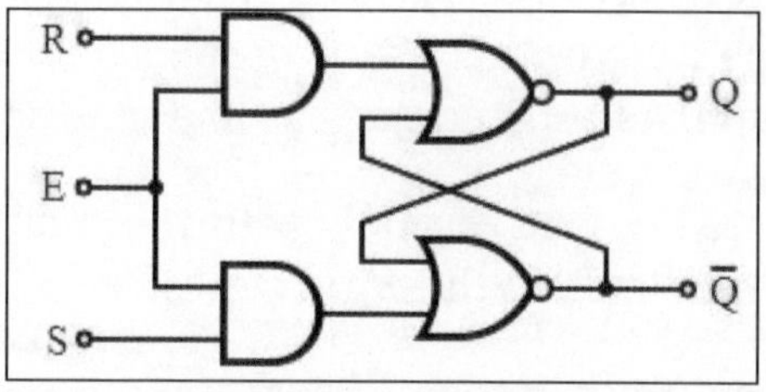

A gated SR latch circuit diagram constructed from AND gates (on left) and NOR gates (on right).

A *synchronous SR latch* (sometimes *clocked SR flip-flop*) can be made by adding a second level of NAND gates to the inverted SR latch (or a second level of AND gates to the direct SR latch). The extra NAND gates further invert the inputs so the simple $\overline{SR}$ latch becomes a gated SR latch (and a simple SR latch would transform into a gated $\overline{SR}$ latch with inverted enable).

With E high (*enable* true), the signals can pass through the input gates to the encapsulated latch; all signal combinations except for (0,0) = *hold* then immediately reproduce on the (Q,Q) output, i.e. the latch is *transparent*.

With E low (*enable* false) the latch is *closed (opaque)* and remains in the state it was left the last time E was high.

The *enable* input is sometimes a clock signal, but more often a read or write strobe. When the *enable* input is a clock signal, the latch is said to be level-sensitive (to the level of the clock signal), as opposed to edge-sensitive like flip-flops below.

Gated SR latch operation	
E/C	Action
0	No action (keep state)
1	The same as non-clocked SR latch

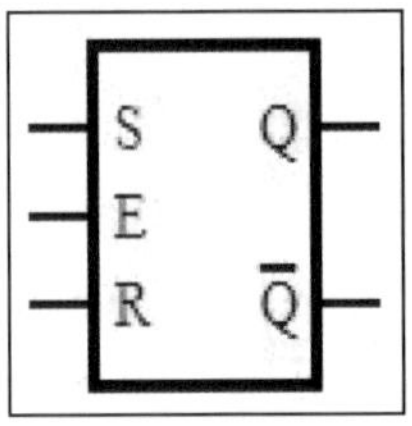

Symbol for a gated SR latch

Gated D Latch

This latch exploits the fact that, in the two active input combinations (01 and 10) of a gated $\overline{SR}$ latch, R is the complement of S. The input NAND stage converts the two D input states (0 and 1) to these two input combinations for the next SR latch by inverting the data input signal. The low state of the *enable* signal produces the inactive "11" combination. Thus a gated D-latch may be considered as a *one-input synchronous SR latch*. This configuration prevents application of the restricted input combination. It is also known as *transparent latch*, *data latch*, or simply *gated latch*. It has a *data* input and an *enable* signal (sometimes named *clock*, or *control*). The word *transparent* comes from the fact that, when the enable input is on, the signal propagates directly through the circuit, from the input D to the output Q. Gated D-latches are also level-sensitive with respect to the level of the clock or enable signal.

Transparent latches are typically used as I/O ports or in asynchronous systems, or in synchronous two-phase systems (synchronous systems that use a two-phase clock), where two latches operating on different clock phases prevent data transparency as in a master–slave flip-flop.

Latches are available as integrated circuits, usually with multiple latches per chip. For example, 74HC75 is a quadruple transparent latch in the 7400 series.

The truth table below shows that when the enable/clock input is 0, the D input has no effect on the output. When E/C is high, the output equals D.

Gated D latch truth table				
E/C	D	Q	$\overline{Q}$	Comment
0	X	Q_{prev}	$\overline{Q}_{prev}$	No change
1	0	0	1	Reset
1	1	1	0	Set

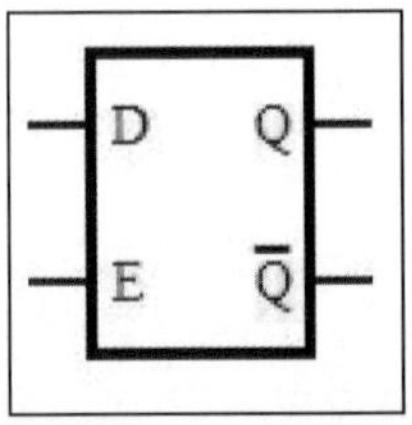

Symbol for a gated D latch

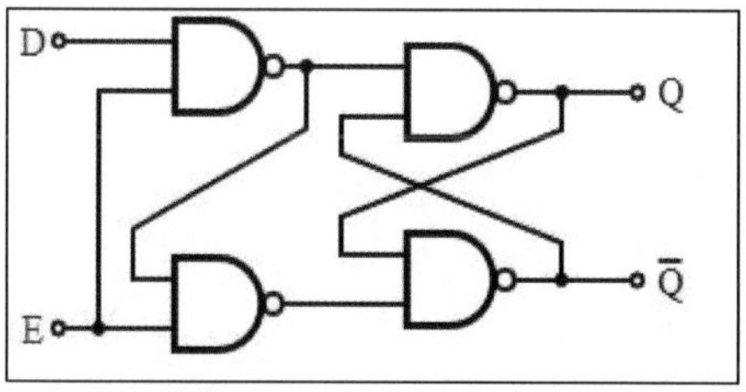

A gated D latch based on an $\overline{SR}$ NAND latch.

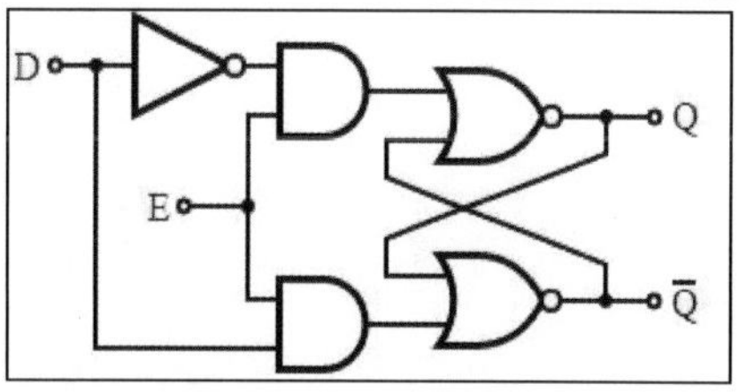

A gated D latch based on an SR NOR latch.

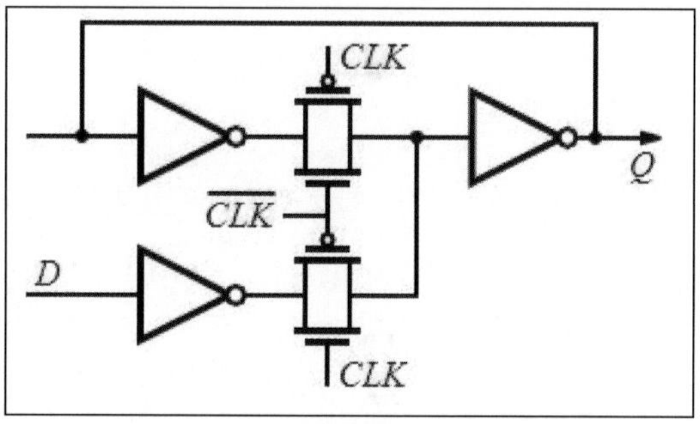

A gated D latch in pass transistor logic, similar to the ones in the CD4042 or the CD74HC75 integrated circuits.

Earle Latch

The classic gated latch designs have some undesirable characteristics. They require double-rail logic or an inverter. The input-to-output propagation may take up to three gate delays. The input-to-output propagation is not constant – some outputs take two gate delays while others take three.

Designers looked for alternatives. A successful alternative is the Earle latch. It requires only a single data input, and its output takes a constant two gate delays. In addition, the two gate levels of the Earle latch can, in some cases, be merged with the last two gate levels of the circuits driving the latch because many common computational circuits have an OR layer followed by an AND layer as their last two levels. Merging the latch function can implement the latch with no additional gate delays. The merge is commonly exploited in the design of pipelined computers, and, in fact, was originally developed by John G. Earle to be used in the IBM System/360 Model 91 for that purpose.

The Earle latch is hazard free. If the middle NAND gate is omitted, then one gets the polarity hold latch, which is commonly used because it demands less logic. However, it is susceptible to logic hazard. Intentionally skewing the clock signal can avoid the hazard.

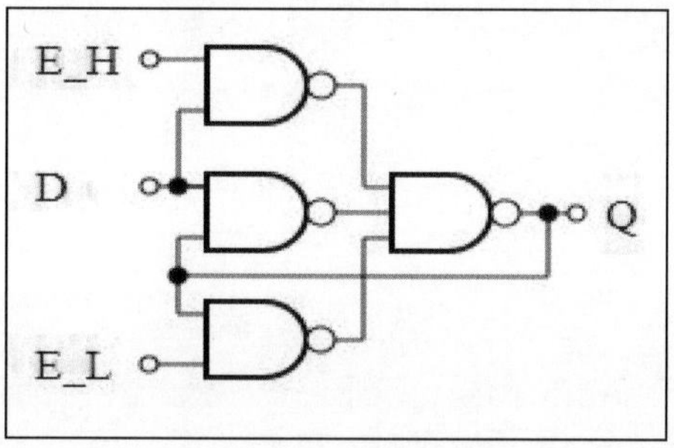

Earle latch uses complementary enable inputs: enable active low (E_L) and enable active high (E_H)

D Flip-flop

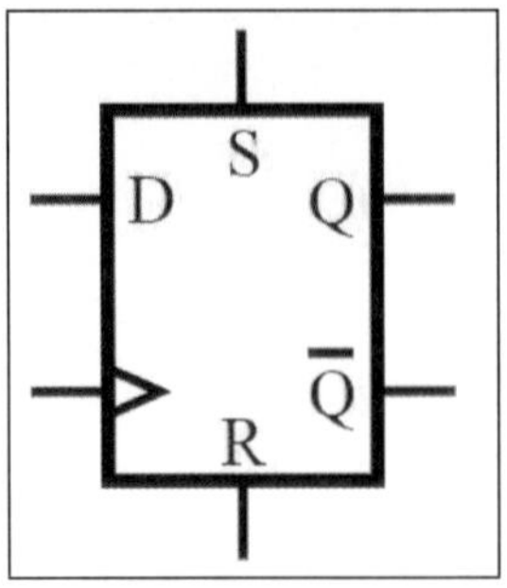

D flip-flop symbol.

The D flip-flop is widely used. It is also known as a “data” or “delay” flip-flop.

The D flip-flop captures the value of the D-input at a definite portion of the clock cycle (such as the rising edge of the clock). That captured value becomes the Q output. At other times, the output Q does not change. The D flip-flop can be viewed as a memory cell, a zero-order hold, or a delay line.

Truth table:

Clock	D	Q_{next}
Rising edge	0	0
Rising edge	1	1
Non-Rising	X	Q

(‘X’ denotes a *Don’t care* condition, meaning the signal is irrelevant).

Most D-type flip-flops in ICs have the capability to be forced to the set or reset state (which ignores the D and clock inputs), much like an SR flip-flop. Usually, the illegal S = R = 1 condition is resolved in D-type flip-flops. Setting S = R = 0 makes the flip-flop behave as described above. Here is the truth table for the others S and R possible configurations:

Inputs				Outputs	
S	R	D	>	Q	Q’
0	1	X	X	0	1
1	0	X	X	1	0
1	1	X	X	1	1

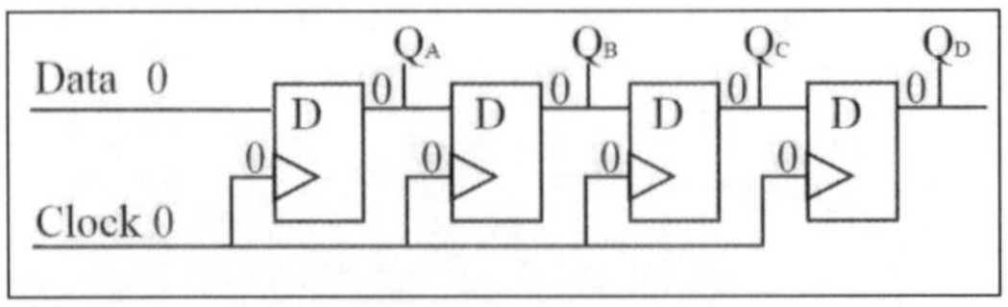

4-bit serial-in, parallel-out (SIPO) shift register.

These flip-flops are very useful, as they form the basis for shift registers, which are an essential part of many electronic devices. The advantage of the D flip-flop over the D-type “transparent latch” is that the signal on the D input pin is captured the moment the flip-flop is clocked, and subsequent changes on the D input will be ignored until the next clock event. An exception is that some flip-flops have a "reset" signal input, which will reset Q (to zero), and may be either asynchronous or synchronous with the clock.

The above circuit shifts the contents of the register to the right, one bit position on each active transition of the clock. The input X is shifted into the leftmost bit position.

Classical Positive-edge-triggered D Flip-flop

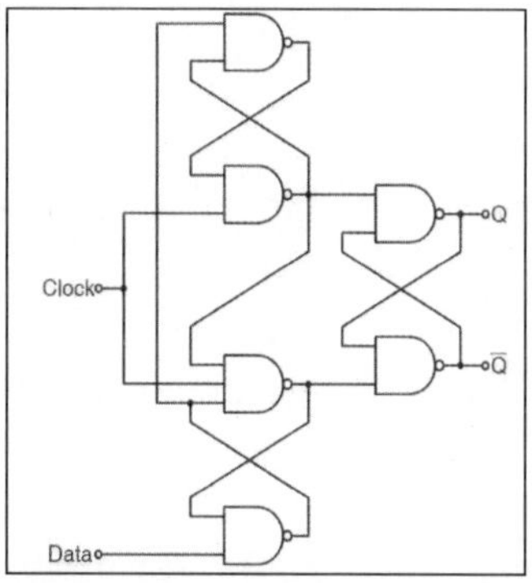

A positive-edge-triggered D flip-flop.

This circuit consists of two stages implemented by SR NAND latches. The input stage (the two latches on the left) processes the clock and data signals to ensure correct input signals for the output stage (the single latch on the right). If the clock is low, both the output signals of the input stage are high regardless of the data input; the output latch is unaffected and it stores the previous state. When the clock signal changes from low to high, only one of the output voltages (depending on the data signal) goes low and sets/resets the output latch: if D = 0, the lower output becomes low; if D = 1, the upper output becomes low. If the clock signal continues staying high, the outputs keep their states regardless of the data input and force the output latch to stay in the corresponding state as the input logical zero (of the output stage) remains active while the clock is high. Hence the role of the output latch is to store the data only while the clock is low.

The circuit is closely related to the gated D latch as both the circuits convert the two D input states (0 and 1) to two input combinations (01 and 10) for the output SR latch by inverting the data input signal (both the circuits split the single D signal in two complementary S and R signals). The difference is that in the gated D latch simple NAND logical gates are used while in the positive-edge-triggered D flip-flop SR NAND latches are used for this purpose. The role of these latches is to "lock" the active output producing low voltage (a logical zero); thus the positive-edge-triggered D flip-flop can also be thought of as a gated D latch with latched input gates.

Master–slave edge-triggered D Flip-flop

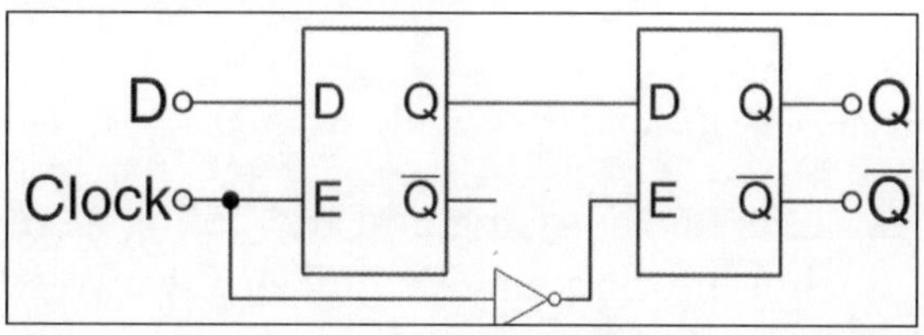

A master–slave D flip-flop. It responds on the falling edge of the *enable* input (usually a clock).

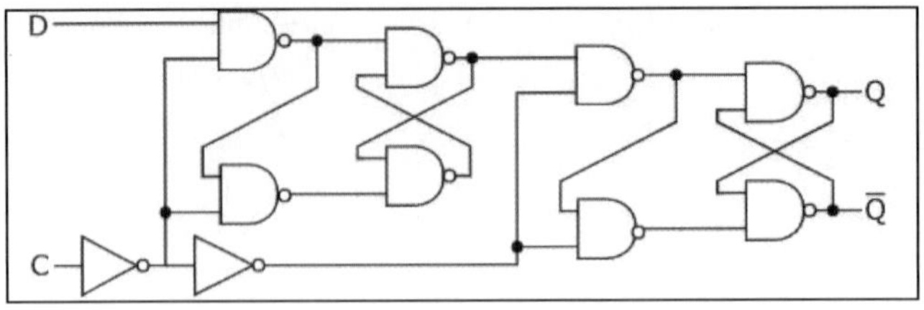

An implementation of a master–slave D flip-flop that is triggered on the rising edge of the clock.

A master–slave D flip-flop is created by connecting two gated D latches in series, and inverting the *enable* input to one of them. It is called master–slave because the second latch in the series changes only in response to a change in the first (master) latch.

For a positive-edge triggered master–slave D flip-flop, when the clock signal is low (logical 0) the "enable" seen by the first or "master" D latch (the inverted clock signal) is high (logical 1). This allows the "master" latch to store the input value when the clock signal transitions from low to high. As the clock signal goes high (0 to 1) the inverted "enable" of the first latch goes low (1 to 0) and the value seen at the input to the master latch is "locked". Nearly simultaneously, the twice inverted "enable" of the second or "slave" D latch transitions from low to high (0 to 1) with the clock signal. This allows the signal captured at the rising edge of the clock by the now "locked" master latch to pass through the "slave" latch. When the clock signal returns to low (1 to 0), the output of the "slave" latch is "locked", and the value seen at the last rising edge of the clock is held while the "master" latch begins to accept new values in preparation for the next rising clock edge.

Removing the leftmost inverter in the circuit creates a D-type flip-flop that strobes on the *falling edge* of a clock signal. This has a truth table like this:

D	Q	>	Q_{next}
0	X	Falling	0
1	X	Falling	1

Dual-edge-triggered D Flip-flop

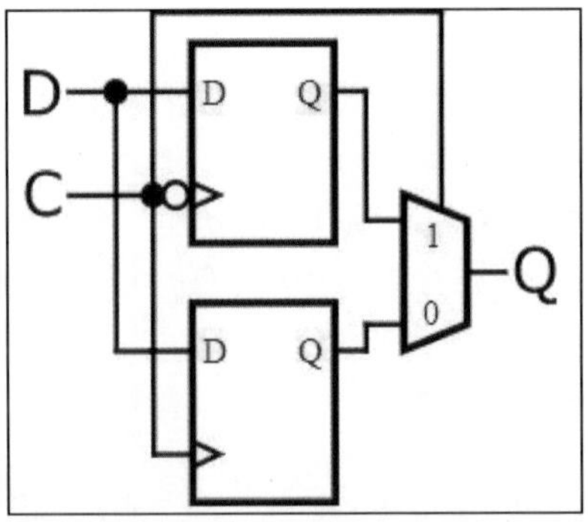

An implementation of a dual-edge-triggered D flip-flop.

Flip-Flops that read in a new value on the rising and the falling edge of the clock are called dual-edge-triggered flip-flops. Such a flip-flop may be built using two single-edge-triggered D-type flip-flops and a multiplexer as shown in the image.

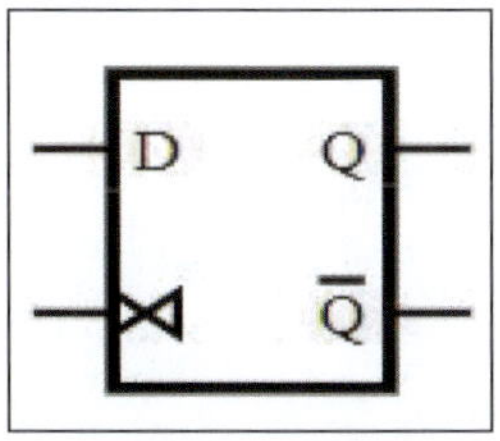

Circuit symbol of a dual-edge-triggered D flip-flop.

Edge-triggered Dynamic D Storage Element

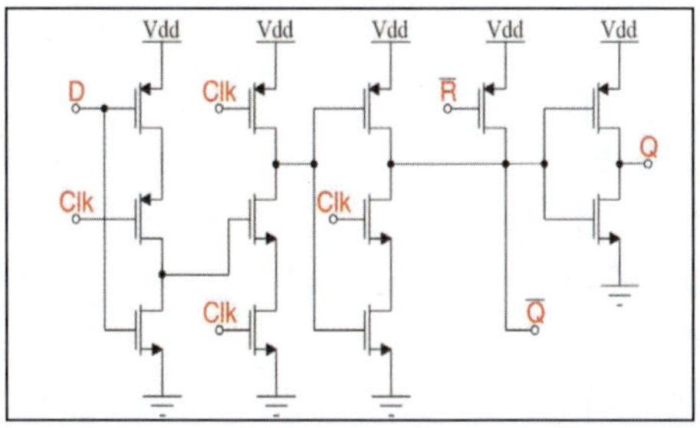

A CMOS IC implementation of a dynamic edge-triggered flip-flop with reset.

An efficient functional alternative to a D flip-flop can be made with dynamic circuits (where information is stored in a capacitance) as long as it is clocked often enough; while not a true flip-flop, it is still called a flip-flop for its functional role. While the master–slave D element is triggered on the edge of a clock, its components are each triggered by clock levels. The "edge-triggered D flip-flop", as it is called even though it is not a true flip-flop, does not have the master–slave properties.

Edge-triggered D flip-flops are often implemented in integrated high-speed operations using dynamic logic. This means that the digital output is stored on parasitic device capacitance while the device is not transitioning. This design of dynamic flip flops also enables simple resetting since the reset operation can be performed by simply discharging one or more internal nodes. A common dynamic flip-flop variety is the true single-phase clock (TSPC) type which performs the flip-flop operation with little power and at high speeds. However, dynamic flip-flops will typically not work at static or low clock speeds: given enough time, leakage paths may discharge the parasitic capacitance enough to cause the flip-flop to enter invalid states.

T Flip-flop

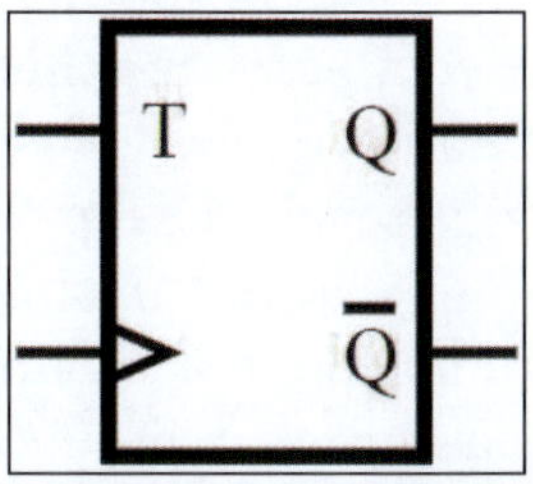

A circuit symbol for a T-type flip-flop.

If the T input is high, the T flip-flop changes state ("toggles") whenever the clock input is strobed. If the T input is low, the flip-flop holds the previous value. This behavior is described by the characteristic equation:

$$Q_{next} = T \oplus Q = T\overline{Q} + \overline{T}Q \text{ (expanding the XOR operator)}$$

and can be described in a truth table:

T flip-flop operation							
Characteristic table				Excitation table			
T	Q	Q_{next}	Comment	Q	Q_{next}	T	Comment
0	0	0	hold state (no clk)	0	0	0	No change
0	1	1	hold state (no clk)	1	1	0	No change
1	0	1	toggle	0	1	1	Complement
1	1	0	toggle	1	0	1	Complement

When T is held high, the toggle flip-flop divides the clock frequency by two; that is, if clock frequency is 4 MHz, the output frequency obtained from the flip-flop will be 2 MHz. This "divide by" feature has application in various types of digital counters. A T flip-flop can also be built using a JK flip-flop (J & K pins are connected together and act as T) or a D flip-flop (T input XOR $Q_{previous}$ drives the D input).

JK Flip-flop

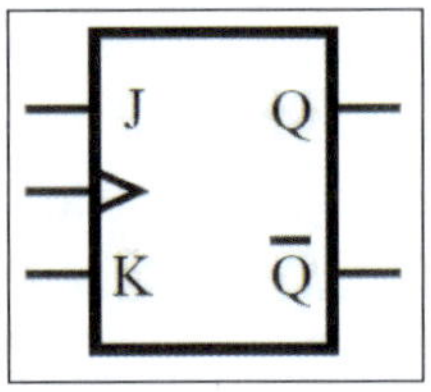

A circuit symbol for a positive-edge-triggered JK flip-flop.

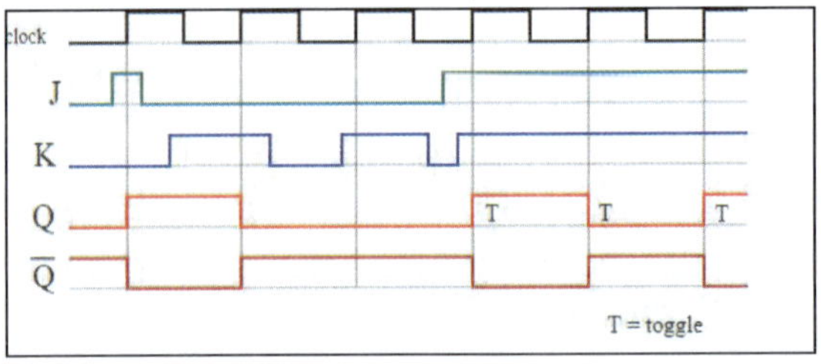

JK flip-flop timing diagram.

The JK flip-flop augments the behavior of the SR flip-flop (J=Set, K=Reset) by interpreting the J = K = 1 condition as a "flip" or toggle command. Specifically, the combination J = 1, K = 0 is a command to set the flip-flop; the combination J = 0, K = 1 is a command to reset the flip-flop; and the combination J = K = 1 is a command to toggle the flip-flop, i.e., change its output to the logical complement of its current value. Setting J = K = 0 maintains the current state. To synthesize a D flip-flop, simply set K equal to the complement of J. Similarly, to synthesize a T flip-flop, set K equal to J. The JK flip-flop is therefore a universal flip-flop, because it can be configured to work as an SR flip-flop, a D flip-flop, or a T flip-flop.

The characteristic equation of the JK flip-flop is:

$$Q_{next} = J\overline{Q} + \overline{K}Q$$

and the corresponding truth table is:

JK flip-flop operation								
Characteristic table				Excitation table				
J	K	Comment	Q_{next}	Q	Q_{next}	Comment	J	K
0	0	hold state	Q	0	0	No Change	0	X
0	1	reset	0	0	1	Set	1	X
1	0	set	1	1	0	Reset	X	1
1	1	toggle	Q	1	1	No Change	X	0

Timing Considerations

Timing Parameters

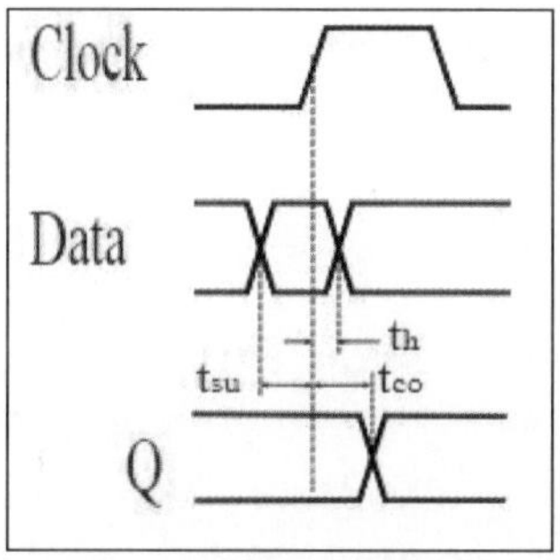

Flip-flop setup, hold and clock-to-output timing parameters.

The input must be held steady in a period around the rising edge of the clock known as the aperture. Imagine taking a picture of a frog on a lily-pad. Suppose the frog then jumps into the water. If you take a picture of the frog as it jumps into the water, you will get a blurry picture of the frog jumping into the water—it's not clear which state the frog was in. But if you take a picture while the frog sits steadily on the pad (or is steadily in the water), you will get a clear picture. In the same way, the input to a flip-flop must be held steady during the aperture of the flip-flop.

Setup time is the minimum amount of time the data input should be held steady before the clock event, so that the data is reliably sampled by the clock.

Hold time is the minimum amount of time the data input should be held steady after the clock event, so that the data is reliably sampled by the clock.

Aperture is the sum of setup and hold time. The data input should be held steady throughout this time period.

Recovery time is the minimum amount of time the asynchronous set or reset input should be inactive before the clock event, so that the data is reliably sampled by the clock. The recovery time for the asynchronous set or reset input is thereby similar to the setup time for the data input.

Removal time is the minimum amount of time the asynchronous set or reset input should be inactive after the clock event, so that the data is reliably sampled by the clock. The removal time for the asynchronous set or reset input is thereby similar to the hold time for the data input.

Short impulses applied to asynchronous inputs (set, reset) should not be applied completely within the recovery-removal period, or else it becomes entirely indeterminable whether the flip-flop will transition to the appropriate state. In another case, where an asynchronous signal simply makes one transition that happens to fall between the recovery/removal time, eventually the flip-flop will transition to the appropriate state, but a very short glitch may or may not appear on the output, dependent on the synchronous input signal. This second situation may or may not have significance to a circuit design.

Set and Reset (and other) signals may be either synchronous or asynchronous and therefore may be characterized with either Setup/Hold or Recovery/Removal times, and synchronicity is very dependent on the design of the flip-flop.

Differentiation between Setup/Hold and Recovery/Removal times is often necessary when verifying the timing of larger circuits because asynchronous signals may be found to be less critical than synchronous signals. The differentiation offers circuit designers the ability to define the verification conditions for these types of signals independently.

Metastability

Flip-flops are subject to a problem called metastability, which can happen when two inputs, such as data and clock or clock and reset, are changing at about the same time. When the order is not clear, within appropriate timing constraints, the result is that the output may behave unpredictably, taking many times longer than normal to settle to one state or the other, or even oscillating several times before settling. Theoretically, the time to settle down is not bounded. In a computer system, this metastability can cause corruption of data or a program crash if the state is not stable before another circuit uses its value; in particular, if two different logical paths use the output of a flip-flop, one path can interpret it as a 0 and the other as a 1 when it has not resolved to stable state, putting the machine into an inconsistent state.

The metastability in flip-flops can be avoided by ensuring that the data and control inputs are held valid and constant for specified periods before and after the clock pulse, called the setup time (t_{su}) and the hold time (t_h) respectively. These times are specified in the data sheet for the device, and are typically between a few nanoseconds and a few hundred picoseconds for modern devices. Depending upon the flip-flop's internal organization, it is possible to build a device with a zero (or even negative) setup or hold time requirement but not both simultaneously.

Unfortunately, it is not always possible to meet the setup and hold criteria, because the flip-flop may be connected to a real-time signal that could change at any time, outside the control of the designer. In this case, the best the designer can do is to reduce the probability of error to a certain level, depending on the required reliability of the circuit. One technique for suppressing metastability is to connect two or more flip-flops in a chain, so that the output of each one feeds the data input of the next, and all devices share a common clock. With this method, the probability of a metastable event can be reduced to a negligible value, but never to zero. The probability of metastability gets closer and closer to zero as the number of flip-flops connected in series is increased. The number of flip-flops being cascaded is referred to as the "ranking"; "dual-ranked" flip flops (two flip-flops in series) is a common situation.

So-called metastable-hardened flip-flops are available, which work by reducing the setup and hold times as much as possible, but even these cannot eliminate the problem entirely. This is because metastability is more than simply a matter of circuit design. When the transitions in the clock and the data are close together in time, the flip-flop is forced to decide which event happened first. However fast the device is made, there is always the possibility that the input events will be so close together that it cannot detect which one happened first. It is therefore logically impossible to build a perfectly metastable-proof flip-flop. Flip-flops are sometimes characterized for a maximum settling time (the maximum time they will remain metastable under specified conditions). In this case, dual-ranked flip-flops that are clocked slower than the maximum allowed metastability time will provide proper conditioning for asynchronous (e.g., external) signals.

Propagation Delay

Another important timing value for a flip-flop is the clock-to-output delay (common symbol in data sheets: t_{CO}) or propagation delay (t_P), which is the time a flip-flop takes to change its output after the clock edge. The time for a high-to-low transition (t_{PHL}) is sometimes different from the time for a low-to-high transition (t_{PLH}).

When cascading flip-flops which share the same clock (as in a shift register), it is important to ensure that the t_{CO} of a preceding flip-flop is longer than the hold time (t_h) of the following flip-flop, so data present at the input of the succeeding flip-flop is properly "shifted in" following the active edge of the clock. This relationship between t_{CO} and t_h is normally guaranteed if the flip-flops are physically identical. Furthermore, for correct operation, it is easy to verify that the clock period has to be greater than the sum $t_{su} + t_h$.

Generalizations

Flip-flops can be generalized in at least two ways: by making them 1-of-N instead of 1-of-2, and by adapting them to logic with more than two states. In the special cases of 1-of-3 encoding, or multi-valued ternary logic, these elements may be referred to as *flip-flap-flops*.

In a conventional flip-flop, exactly one of the two complementary outputs is high. This can be generalized to a memory element with N outputs, exactly one of which is high (alternatively, where exactly one of N is low). The output is therefore always a one-hot (respectively *one-cold*) representation. The construction is similar to a conventional cross-coupled flip-flop; each output, when high, inhibits all the other outputs. Alternatively, more or less conventional flip-flops can be used, one per output, with additional circuitry to make sure only one at a time can be true.

Another generalization of the conventional flip-flop is a memory element for multi-valued logic. In this case the memory element retains exactly one of the logic states until the control inputs induce a change. In addition, a multiple-valued clock can also be used, leading to new possible clock transitions.

Shift Register

In digital circuits, a shift register is a cascade of flip flops, sharing the same clock, in which the output of each flip-flop is connected to the "data" input of the next flip-flop in the chain, resulting in a circuit that shifts by one position the "bit array" stored in it, "shifting in" the data present at its input and 'shifting out' the last bit in the array, at each transition of the clock input.

More generally, a shift register may be multidimensional, such that its "data in" and stage outputs are themselves bit arrays; this is implemented simply by running several shift registers of the same bit-length in parallel.

Shift registers can have both parallel and serial inputs and outputs. These are often configured as "serial-in, parallel-out" (SIPO) or as "parallel-in, serial-out" (PISO). There are also types that have both serial and parallel input and types with serial and parallel output. There are also "bidirectional" shift registers which allow shifting in both directions: L→R or R→L. The serial input and last output of a shift register can also be connected to create a "circular shift register". One register is PIPO (parallel in parallel out), which is very fast, within single clock pulse, it is giving output.

Serial-in Serial-out (SISO)

Destructive Readout

0	0	0	0
1	0	0	0
0	1	0	0
1	0	1	0
1	1	0	1
0	1	1	0
0	0	1	1
0	0	0	1
0	0	0	0

These are the simplest kind of shift registers. The data string is presented at "Data In", and is shifted right one stage each time "Data Advance" is brought high. At each advance, the bit on the far left (i.e. "Data In") is shifted into the first flip-flop's output. The bit on the far right (i.e. "Data Out") is shifted out and lost.

The data is stored after each flip-flop on the "Q" output, so there are four storage "slots" available in this arrangement, hence it is a 4-bit Register. To give an idea of the shifting pattern, imagine that the register holds 0000 (so all storage slots are empty). As "Data In" presents 1,0,1,1,0,0,0,0 (in that order, with a pulse at "Data Advance" each time—this is called clocking or strobing) to the register, this is the result. The right hand column corresponds to the right-most flip-flop's output pin, and so on.

So the serial output of the entire register is 00001011. It can be seen that if data were to be continued to input, it would get exactly what was put in (10110000), but offset by four "Data Advance" cycles. This arrangement is the hardware equivalent of a queue. Also, at any time, the whole register can be set to zero by bringing the reset (R) pins high.

This arrangement performs *destructive readout* - each datum is lost once it has been shifted out of the right-most bit.

Serial-In Parallel-Out (SIPO)

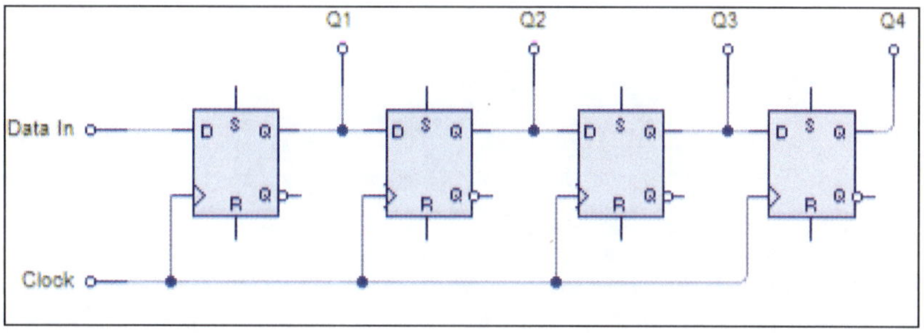

This configuration allows conversion from serial to parallel format. Data input is serial, as described in the SISO section above. Once the data has been clocked in, it may be either read off at each output simultaneously, or it can be shifted out.

In this configuration, each flip-flop is edge triggered. All flip-flops operate at the given clock frequency. Each input bit makes its way down to the Nth output after N clock cycles, leading to parallel output.

In cases where the parallel outputs should not change during the serial loading process, it is desirable to use a latched or buffered output. In a latched shift register (such as the 74595) the serial data is first loaded into an internal buffer register, then upon receipt of a load signal the state of the buffer register is copied into a set of output registers. In general, the practical application of the serial-in/parallel-out shift register is to convert data from serial format on a single wire to parallel format on multiple wires.

Parallel-In Serial-Out (PISO)

This configuration has the data input on lines D1 through D4 in parallel format, D1 being the most significant bit. To write the data to the register, the Write/Shift control line must be held LOW. To shift the data, the W/S control line is brought HIGH and the registers are clocked. The arrangement now acts as a PISO shift register, with D1 as the Data Input. However, as long as the number of clock cycles is not more than the length of the data-string, the Data Output, Q, will be the parallel data read off in order.

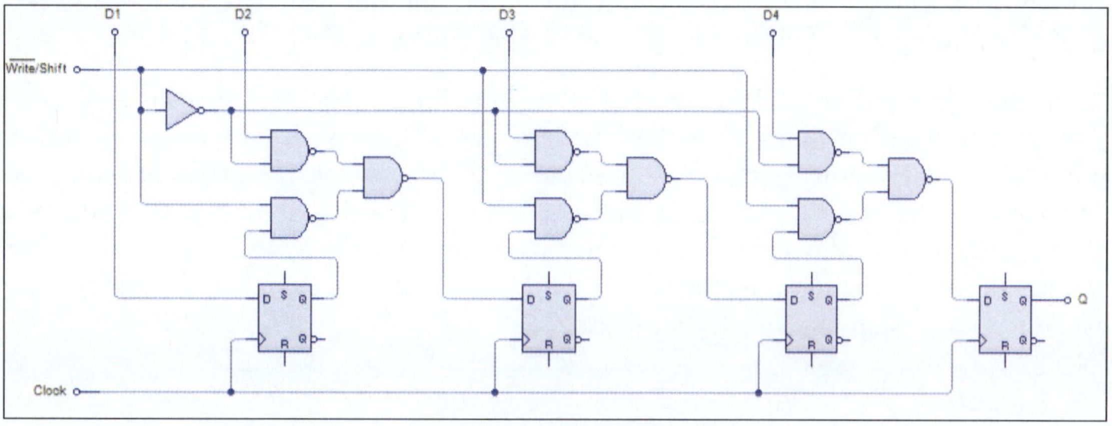

4-Bit PISO Shift Register.

Uses

Toshiba TC4015BP - Dual 4-Stage Static Shift Register (with serial input/parallel output).

One of the most common uses of a shift register is to convert between serial and parallel interfaces. This is useful as many circuits work on groups of bits in parallel, but serial interfaces are simpler to construct. Shift registers can be used as simple delay circuits. Several bidirectional shift registers could also be connected in parallel for a hardware implementation of a stack.

SIPO registers are commonly attached to the output of microprocessors when more General Purpose Input/Output pins are required than are available. This allows several binary devices to be controlled using only two or three pins, but slower than parallel I/O - the devices in question are attached to the parallel outputs of the shift register, then the desired state of all those devices can be sent out of the microprocessor using a single serial connection. Similarly, PISO configurations are commonly used to add more binary inputs to a microprocessor than are available - each binary input (i.e. a button or more complicated circuitry) is attached to a parallel input of the shift register, then the data is sent back via serial to the microprocessor using several fewer lines than originally required.

Shift registers can also be used as pulse extenders. Compared to monostable multivibrators, the timing has no dependency on component values, however, it requires external clock and the timing accuracy is limited by a granularity of this clock. Example: Ronja Twister, where five 74164 shift registers create the core of the timing logic this way (schematic).

In early computers, shift registers were used to handle data processing: two numbers to be added were stored in two shift registers and clocked out into an arithmetic and logic unit (ALU) with the result being fed back to the input of one of the shift registers (the accumulator) which was one bit longer since binary addition can only result in an answer that is the same size or one bit longer.

Many computer languages include instructions to "shift right" and "shift left" the data in a register, effectively dividing by two or multiplying by two for each place shifted.

Very large serial-in serial-out shift registers (thousands of bits in size) were used in a similar manner to the earlier delay line memory in some devices built in the early 1970s. Such memories were sometimes called "circulating memory". For example, the Datapoint 3300 terminal stored its display of 25 rows of 72 columns of upper-case characters using fifty-four 200-bit shift registers, arranged in six tracks of nine packs each, providing storage for 1800 six-bit characters. The shift register design meant that scrolling the terminal display could be accomplished by simply pausing the display output to skip one line of characters.

Digital Counters

Counter is a sequential circuit. A digital circuit which is used for a counting pulses is known counter. Counter is the widest application of flip-flops. It is a group of flip-flops with a clock signal applied. Counters are of two types.

- Asynchronous or ripple counters.
- Synchronous counters.

Asynchronous or Ripple Counters

The logic diagram of a 2-bit ripple up counter is shown in figure. The toggle (T) flip-flop are being used. But we can use the JK flip-flop also with J and K connected permanently to logic 1. External clock is applied to the clock input of flip-flop A and QA output is applied to the clock input of the next flip-flop i.e. FF-B.

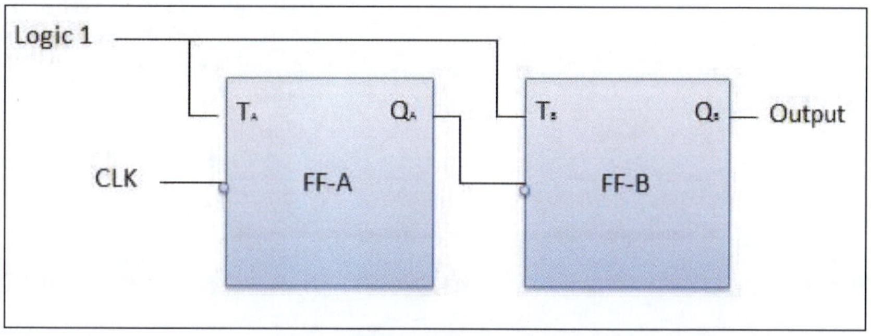

Logical Diagram.

Operation

S.N.	Condition	Operation
1	Initially let both the FFs be in the reset state	Q_BQ_A = 00 initially.
2	After 1st negative clock edge	As soon as the first negative clock edge is applied, FF-A will toggle and Q_A will be equal to 1. Q_A is connected to clock input of FF-B. Since Q_A has changed from 0 to 1, it is treated as the positive clock edge by FF-B. There is no change in Q_B because FF-B is a negative edge triggered FF. Q_BQ_A = 01 after the first clock pulse.
3	After 2nd negative clock edge	On the arrival of second negative clock edge, FF-A toggles again and Q_A = 0. The change in Q_A acts as a negative clock edge for FF-B. So it will also toggle, and Q_B will be 1. Q_BQ_A = 10 after the second clock pulse.

4	After 3rd negative clock edge	On the arrival of 3rd negative clock edge, FF-A toggles again and Q_A become 1 from 0. Since this is a positive going change, FF-B does not respond to it and remains inactive. So Q_B does not change and continues to be equal to 1. Q_BQ_A = 11 after the third clock pulse.
5	After 4th negative clock edge	On the arrival of 4th negative clock edge, FF-A toggles again and Q_A becomes 1 from 0. This negative change in Q_A acts as clock pulse for FF-B. Hence it toggles to change Q_B from 1 to 0. Q_BQ_A = 00 after the fourth clock pulse.

Truth Table

Clock	Counter output		State number	Deciimal Counter output
	Q	Q_A		
Initially	0	0	-	0
1st	0	1	1	1
2nd	1	0	2	2
3rd	1	1	3	3
4th	0	0	4	0

Synchronous Counters

If the "clock" pulses are applied to all the flip-flops in a counter simultaneously, then such a counter is called as synchronous counter.

2-bit Synchronous up Counter

The JA and KA inputs of FF-A are tied to logic 1. So FF-A will work as a toggle flip-flop. The JB and KB inputs are connected to QA.

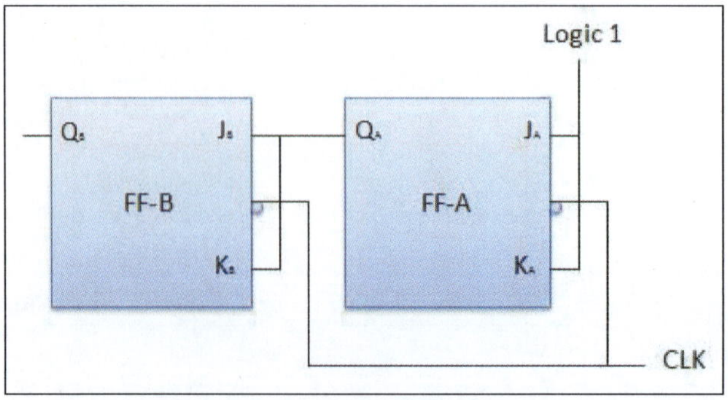

Logical Diagram.

Operation

S.N.	Condition	Operation
1	Initially let both the FFs be in the reset state	Q_BQ_A = 00 initially.
2	After 1st negative clock edge	As soon as the first negative clock edge is applied, FF-A will toggle and Q_A will change from 0 to 1. But at the instant of application of negative clock edge, Q_A , $J_B = K_B = 0$. Hence FF-B will not change its state. So Q_B will remain 0. Q_BQ_A = 01 after the first clock pulse.
3	After 2nd negative clock edge	On the arrival of second negative clock edge, FF-A toggles again and Q_A changes from 1 to 0. But at this instant Q_A was 1. So $J_B = K_B = 1$ and FF-B will toggle. Hence Q_B changes from 0 to 1. Q_BQ_A = 10 after the second clock pulse.
4	After 3rd negative clock edge	On application of the third falling clock edge, FF-A will toggle from 0 to 1 but there is no change of state for FF-B. Q_BQ_A = 11 after the third clock pulse.
5	After 4th negative clock edge	On application of the next clock pulse, Q_A will change from 1 to 0 as Q_B will also change from 1 to 0. Q_BQ_A = 00 after the fourth clock pulse.

Classification of Counters

Depending on the way in which the counting progresses, the synchronous or asynchronous counters are classified as follows:

- Up counters;
- Down counters;
- Up/Down counters.

UP/DOWN Counter

Up counter and down counter is combined together to obtain an UP/DOWN counter. A mode control (M) input is also provided to select either up or down mode. A combinational circuit is required to be designed and used between each pair of flip-flop in order to achieve the up/down operation.

- Type of up/down counters;
- UP/DOWN ripple counters;

- UP/DOWN synchronous counter.

UP/DOWN Ripple Counters

In the UP/DOWN ripple counter all the FFs operate in the toggle mode. So either T flip-flops or JK flip-flops are to be used. The LSB flip-flop receives clock directly. But the clock to every other FF is obtained from (Q = Q bar) output of the previous FF.

- UP counting mode (M=0) – The Q output of the preceding FF is connected to the clock of the next stage if up counting is to be achieved. For this mode, the mode select input M is at logic 0 (M=0).
- DOWN counting mode (M=1) – If M = 1, then the Q bar output of the preceding FF is connected to the next FF. This will operate the counter in the counting mode.

3-bit binary up/down ripple counter.

- 3-bit – hence three FFs are required.
- UP/DOWN – So a mode control input is essential.
- For a ripple up counter, the Q output of preceding FF is connected to the clock input of the next one.
- For a ripple down counter, the Q bar output of preceding FF is connected to the clock input of the next one.
- Let the selection of Q and Q bar output of the preceding FF be controlled by the mode control input M such that, If M = 0, UP counting. So connect Q to CLK. If M = 1, DOWN counting. So connect Q bar to CLK.

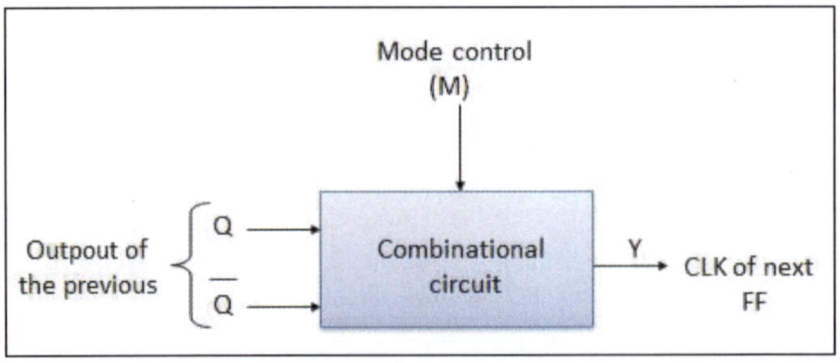

Block Diagram.

Inputs			Outputs	
M	Q	$\overline{Q}$	Y	
0	0	0	0	Y = Q for up counter
0	0	1	0	
0	1	0	1	
0	1	1	1	
1	0	0	0	Y = $\overline{Q}$ for up counter
1	0	1	1	
1	1	0	0	
1	1	1	1	

Truth Table.

Operation

S.N.	Condition	Operation
1	Case 1 – With M = 0 (Up counting mode)	If M = 0 and M bar = 1, then the AND gates 1 and 3 in fig. will be enabled whereas the AND gates 2 and 4 will be disabled. Hence Q_A gets connected to the clock input of FF-B and Q_B gets connected to the clock input of FF-C. These connections are same as those for the normal up counter. Thus with M = 0 the circuit work as an up counter.
2	Case 2: With M = 1 (Down counting mode)	If M = 1, then AND gates 2 and 4 in fig. are enabled whereas the AND gates 1 and 3 are disabled. Hence Q_A bar gets connected to the clock input of FF-B and Q_B bar gets connected to the clock input of FF-C. These connections will produce a down counter. Thus with M = 1 the circuit works as a down counter.

Modulus Counter (MOD-N Counter)

The 2-bit ripple counter is called as MOD-4 counter and 3-bit ripple counter is called as MOD-8 counter. So in general, an n-bit ripple counter is called as modulo-N counter. Where, MOD number = 2^n.

Type of Modulus

- 2-bit up or down (MOD-4);
- 3-bit up or down (MOD-8);
- 4-bit up or down (MOD-16).

Application of Counters

- Frequency counters;
- Digital clock;
- Time measurement;
- A to D converter;
- Frequency divider circuits;
- Digital triangular wave generator.

References

- Digital-electronics, wiki: fandom.com, Retrieved 30 January, 2019

- Michael Predko and Myke Predko, Digital electronics demystified, mcgraw-Hill, 2004. ISBN 0-07-144141-7
- Logic-10, logic: electronics-tutorials.ws, Retrieved 29 June, 2019
- Wolfram, Stephen (2002). A New Kind of Science. Wolfram Media, Inc. P. 1097. ISBN 1-57955-008-8
- Boolean-algebra, computer-logical-organization: tutorialspoint.com, Retrieved 13 March, 2019
- Cook, Aaron. "Using Karnaugh Maps to Simplify Code". Quantum Rarity. Archived from the original on 2017-04-18. Retrieved 2012-10-07
- Comb-3, combination: electronics-tutorials.ws, Retrieved 23 February, 2019
- Horowitz P., Hill W. - The Art of Electronics. 32 Avenue of the Americas, New York, NY 10013-2473, USA. 2015. P. 764. ISBN 978-0-521-80926-9
- Electronics-basics-what-is-a-latch-circuit, diy-projects, electronics, programming: dummies.com, Retrieved 22 May, 2019
- Digital-counters, computer-logical-organization: tutorialspoint.com, Retrieved 2 June, 2019
- "DL-3422 4-digit 22-segment alphanumeric Intelligent Display™ preliminary data sheet". Internet Archive. Litronix 1982 Optoelectronics Catalog. P. 82. Retrieved 2016-09-03

4

Semiconductors

The material that has an electrical conductivity value falling between that of a conductor and an insulator is referred to as a semiconductor. The main types of semiconductor are n-type semiconductors and p-type semiconductors. The chapter closely examines these key types of semiconductors to provide an extensive understanding of the subject.

Semiconductor is any of a class of crystalline solids intermediate in electrical conductivity between a conductor and an insulator. Semiconductors are employed in the manufacture of various kinds of electronic devices, including diodes, transistors, and integrated circuits. Such devices have found wide application because of their compactness, reliability, power efficiency, and low cost. As discrete components, they have found use in power devices, optical sensors, and light emitters, including solid-state lasers. They have a wide range of current- and voltage-handling capabilities and, more important, lend themselves to integration into complex but readily manufacturable microelectronic circuits. They are, and will be in the foreseeable future, the key elements for the majority of electronic systems, serving communications, signal processing, computing, and control applications in both the consumer and industrial markets.

Semiconductor Materials

Solid-state materials are commonly grouped into three classes: insulators, semiconductors, and conductors. (At low temperatures some conductors, semiconductors, and insulators may become superconductors.) Insulators, such as fused quartz and glass, have very low conductivities, on the order of 10^{-18} to 10^{-10} siemens per centimetre; and conductors, such as aluminum, have high conductivities, typically from 10^4 to 10^6 siemens per centimetre. The conductivities of semiconductors are between these extremes and are generally sensitive to temperature, illumination, magnetic fields, and minute amounts of impurity atoms. For example, the addition of about 10 atoms of boron (known as a dopant) per million atoms of silicon can increase its electrical conductivity a thousandfold.

The study of semiconductor materials began in the early 19th century. The elemental semiconductors are those composed of single species of atoms, such as silicon (Si), germanium (Ge), and tin (Sn) in column IV and selenium (Se) and tellurium (Te) in column VI of the periodic table. There are, however, numerous compound semiconductors, which are composed of two or more elements. Gallium arsenide (GaAs), for example, is a binary III-V compound, which is a combination of gallium (Ga) from column III and arsenic (As) from column V. Ternary compounds can be formed by elements from three different columns—for instance, mercury indium telluride ($HgIn_2Te_4$), a II-III-VI compound. They also can be formed by elements from two columns, such as aluminum gallium arsenide ($Al_xGa_{1-x}As$), which is a ternary III-V compound, where both Al and Ga are from column III and the subscript x is related to the composition of the two elements from 100 percent Al ($x = 1$) to 100 percent Ga ($x = 0$). Pure silicon is the most important material for integrated circuit applications, and III-V binary and ternary compounds are most significant for light emission.

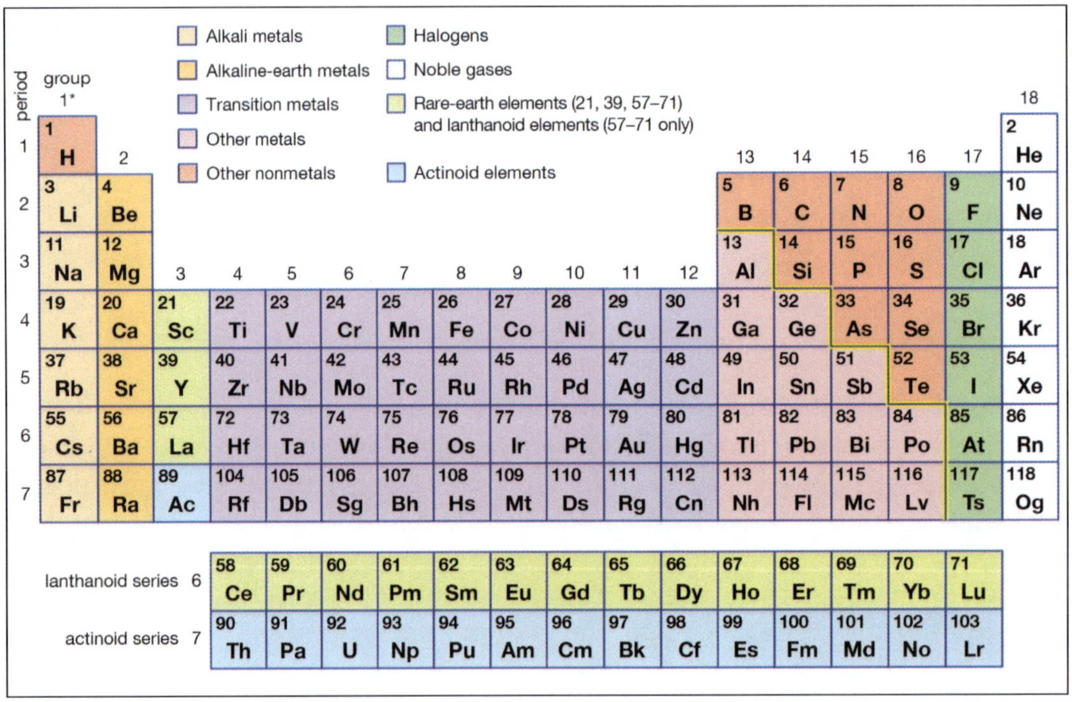

Modern version of the periodic table of the elements.

Prior to the invention of the bipolar transistor in 1947, semiconductors were used only as two-terminal devices, such as rectifiers and photodiodes. During the early 1950s germanium was the major semiconductor material. However, it proved unsuitable for many applications, because devices made of the material exhibited high leakage currents at only moderately elevated temperatures. Since the early 1960s silicon has become by far the most widely used semiconductor, virtually supplanting germanium as a material for device fabrication. The main reasons for this are twofold: (1) silicon devices exhibit much lower leakage currents, and (2) silicon dioxide (SiO_2), which is a high-quality insulator, is easy to incorporate as part of a silicon-based device. Thus, silicon technology has become very advanced and pervasive, with silicon devices constituting more than 95 percent of all semiconductor products sold worldwide.

Many of the compound semiconductors have some specific electrical and optical properties that are superior to their counterparts in silicon. These semiconductors, especially gallium arsenide, are used mainly for optoelectronic and certain radio frequency (RF) applications.

Electronic Properties

The semiconductor materials described here are single crystals; i.e., the atoms are arranged in a three-dimensional periodic fashion. Part A of the figure shows a simplified two-dimensional representation of an intrinsic (pure) silicon crystal that contains negligible impurities. Each silicon atom in the crystal is surrounded by four of its nearest neighbours. Each atom has four electrons in its outer orbit and shares these electrons with its four neighbours. Each shared electron pair constitutes a covalent bond. The force of attraction between the electrons and both nuclei holds the two atoms together. For isolated atoms (e.g., in a gas rather than a crystal), the electrons can have only discrete energy levels. However, when a large number of atoms are brought together to form a crystal, the interaction between the atoms causes the discrete energy levels to spread out into energy bands. When there is no thermal vibration (i.e., at low temperature), the electrons in an insulator or semiconductor crystal will completely fill a number of energy bands, leaving the rest of the energy bands empty. The highest filled band is called the valence band. The next band is the conduction band, which is separated from the valence band by an energy gap (much larger gaps in crystalline insulators than in semiconductors). This energy gap, also called a bandgap, is a region that designates energies that the electrons in the crystal cannot possess. Most of the important semiconductors have bandgaps in the range 0.25 to 2.5 electron volts (eV). The bandgap of silicon, for example, is 1.12 eV, and that of gallium arsenide is 1.42 eV. In contrast, the bandgap of diamond, a good crystalline insulator, is 5.5 eV.

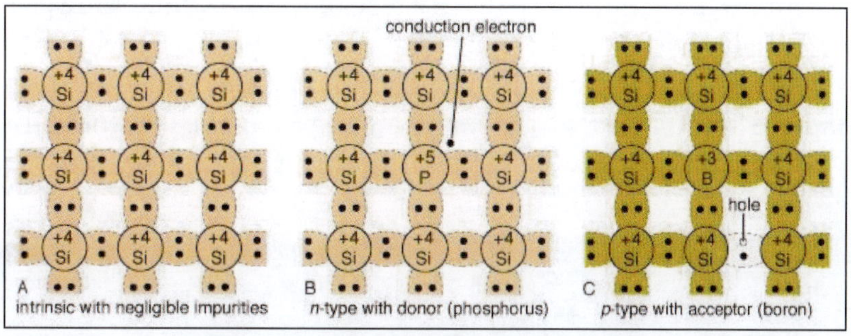

Three bond pictures of a semiconductor.

At low temperatures the electrons in a semiconductor are bound in their respective bands in the crystal; consequently, they are not available for electrical conduction. At higher temperatures thermal vibration may break some of the covalent bonds to yield free electrons that can participate in current conduction. Once an electron moves away from a covalent bond, there is an electron vacancy associated with that bond. This vacancy may be filled by a neighbouring electron, which results in a shift of the vacancy location from one crystal site to another. This vacancy may be regarded as a fictitious particle, dubbed a "hole," that carries a positive charge and moves in a direction opposite to that of an electron. When an electric field is applied to the semiconductor, both the free electrons (now residing in the conduction band) and the holes (left behind in the valence band) move through the crystal, producing an electric current. The electrical conductivity of a material depends on the number of free electrons and holes (charge carriers) per unit volume and on the rate at which these carriers move under the influence of an electric field. In an intrinsic semiconductor there exists an equal number of free electrons and holes. The electrons and holes, however, have different mobilities; that is, they move with different velocities in an electric field. For example, for intrinsic silicon at room temperature, the electron mobility is 1,500 square centimetres per volt-second ($cm^2/V{\cdot}s$)—i.e., an electron will move at a velocity of 1,500 centimetres per second under an electric field of one volt per centimetre—while the hole mobility is 500 $cm^2/V{\cdot}s$. The electron and hole mobilities in a particular semiconductor generally decrease with increasing temperature.

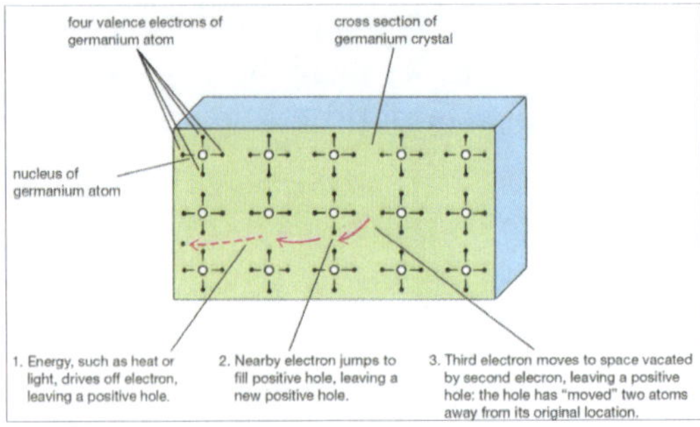

Movement of an electron hole in a crystal lattice.

Electrical conduction in intrinsic semiconductors is quite poor at room temperature. To produce higher conduction, one can intentionally introduce impurities (typically to a concentration of one part per million host atoms). This is called doping, a process that increases conductivity despite some loss of mobility. For example, if a silicon atom is replaced by an atom with five outer electrons, such as arsenic four of the electrons form covalent bonds with the four neighbouring silicon atoms. The fifth electron becomes a conduction electron that is donated to the conduction band. The silicon becomes an n-type semiconductor because of the addition of the electron. The arsenic atom is the donor. Similarly, part C of the figure shows that, if an atom with three outer electrons, such as boron, is substituted for a silicon atom, an additional electron is accepted to form four covalent bonds around the boron atom, and a positively charged hole is created in the valence band. This creates a p-type semiconductor, with the boron constituting an acceptor.

n-Type Semiconductor

When a small amount of Pentavalent impurity is added to a pure semiconductor providing a large number of free electrons in it, the extrinsic semiconductor thus formed is known as n-Type Semiconductor. The conduction in the n-type semiconductor is because of the free electrons denoted by the pentavalent impurity atoms. These electrons are the excess free electrons with regards to the number of free electrons required to fill the covalent bonds in the semiconductors.

The addition of Pentavalent impurities such as arsenic and antimony provides a large number of free electrons in the semiconductor crystal. Such impurities which produce n-type semiconductors are known as Donor Impurities. They are called a donor impurity because each atom of them donates one free electron crystal.

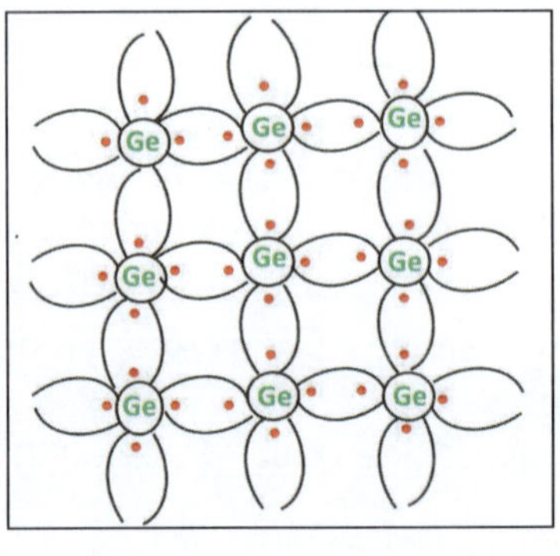

When a few Pentavalent impurities such as Arsenic whose atomic number is 33, which is categorised as 2, 8, 15 and 5. It has five valence electrons, which is added to germanium crystal. Each atom of the impurity fits in four germanium atoms as shown in the figure.

Hence, each Arsenic atom provides one free electron in the Germanium crystal. Since an extremely small amount of arsenic, impurity has a large number of atoms; it provides millions of free electrons for conduction.

Energy Diagram of n-Type Semiconductor

The Energy diagram of the n-type semiconductor is shown in the figure.

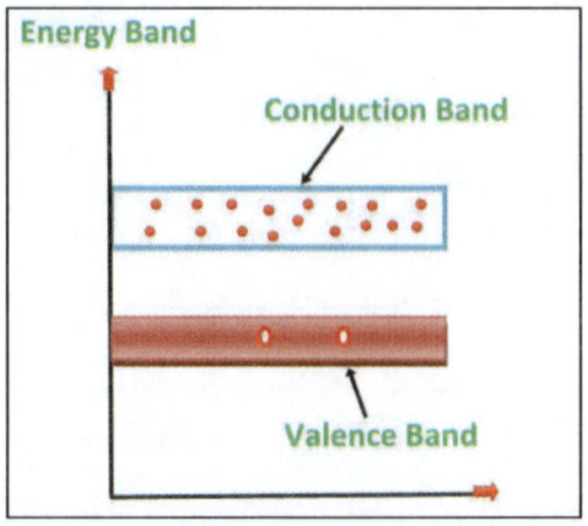

A large number of free electrons are available in the conduction band because of the addition of the Pentavalent impurity. These electrons are free electrons which did not fit in the covalent bonds of the crystal. However, a minute quantity of free electrons is available in the conduction band forming hole- electron pairs.

The following points are important in the n-type semiconductor:

- The addition of Pentavalent impurity results in a large number of free electrons.
- When thermal energy at room temperature is imparted to the semiconductor, a hole-electron pair is generated and as a result, a minute quantity of free electrons are available. These electrons leave behind holes in the valence band.
- Here n stands for negative material as the number of free electrons provided by the Pentavalent impurity is greater than the number of holes.

Conduction through n-Type Semiconductor

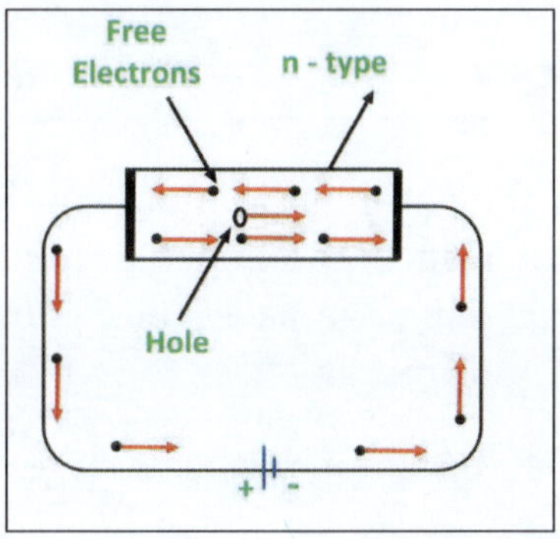

In the n-type semiconductor, a large number of free electrons are available in the conduction band which are donated by the impurity atoms.

When a potential difference is applied across this type of semiconductor, the free electrons are directed towards the positive terminals. It carries an electric current. As the flow of current through the crystal is constituted by free electrons which are carriers of negative charge, therefore, this type of conductivity is known as negative or n-type conductivity.

The electron-hole pairs are formed at room temperature. These holes which are available in small quantity in valence band also consists of a small amount of current. For practical purposes, this current is neglected.

p-Type Semiconductor

The extrinsic p-Type Semiconductor is formed when a trivalent impurity is added to a pure semiconductor in a small amount, and as a result, a large number of holes are created in it. A large number of holes are provided in the semiconductor material by the addition of trivalent impurities like Gallium and Indium. Such type of impurities which produces p-type semiconductor are known as an Acceptor Impurities because each atom of them create one hole which can accept one electron.

A trivalent impurity like gallium, having three valence electrons is added to germanium crystal in a small amount. Each atom of the impurity fits in the germanium crystal in such a way that its three valence electrons form covalent bonds with the three surrounding germanium atoms as shown in the figure.

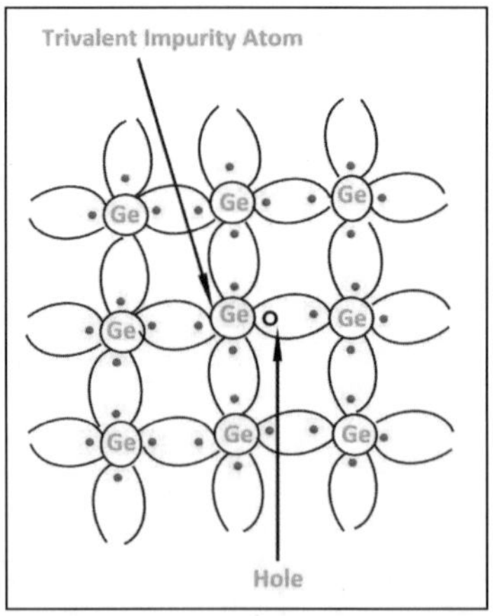

In the fourth covalent bonds, only the germanium atom contributes one valence electron, while gallium atom has no valence bonds. Hence, the fourth covalent bond is incomplete, having one electron short. This missing electron is known as a Hole. Thus, each gallium atom provides one hole in the germanium crystal.

As an extremely small amount of Gallium impurity has a large number of atoms, therefore, it provides millions of holes in the semiconductor.

Energy Band Diagram of p-Type Semiconductor

The energy band diagram of a p-Type Semiconductor is shown.

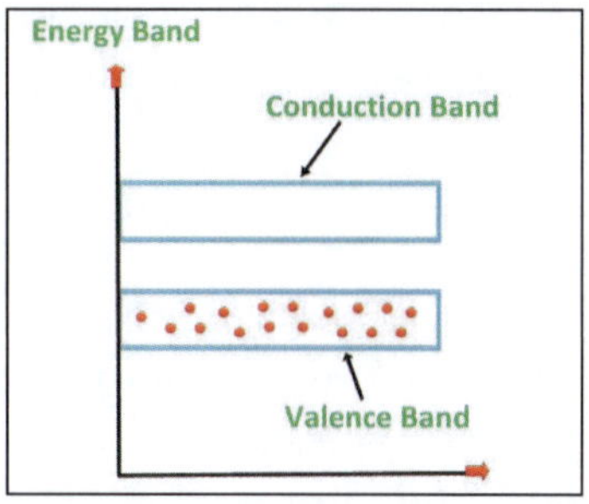

A large number of holes or vacant space in the covalent bond is created in the crystal with the addition of the trivalent impurity. A small or minute quantity of free electrons is also available in the conduction band.

They are produced when thermal energy at room temperature is imparted to the germanium crystal forming electron-hole pairs. But the holes are more in number as compared to the electrons in the conduction band. It is because of the predominance of holes over electrons that the material is called as a p-type semiconductor. The word "p" stands for the positive material.

Conduction through p-Type Semiconductor

In p type semiconductor large number of holes are created by the trivalent impurity. When a potential difference is applied across this type of semiconductor as shown in the figure below.

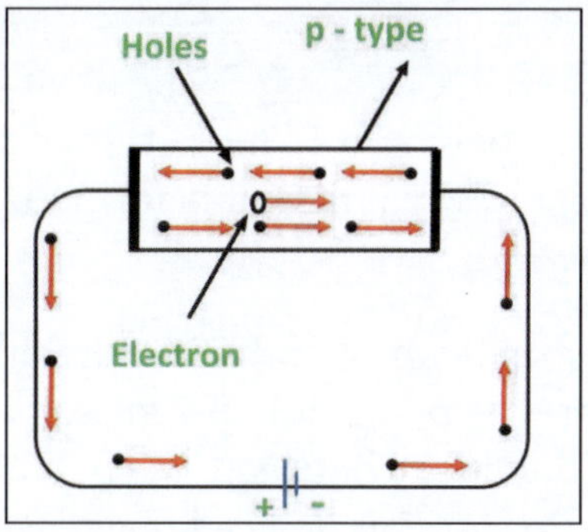

The holes are available in the valence band are directed towards the negative terminal. As the current flow through the crystal is by holes, which are carrier of positive charge, therefore, this type of conductivity is known as positive or p type conductivity. In a p type conductivity the valence electrons move from one covalent to another.

The conductivity of n type semiconductor is nearly double to that of p type semiconductor. The electrons available in the conduction band of the n type semiconductor are much more movable than holes available in the valence band in a p type semiconductor. The mobility of holes is poor as they are more bound to the nucleus.

Even at the room temperature the electron hole pairs are formed. These free electrons which are available in minute quantity also carry a little amount of current in the p type semiconductors.

The p-n Junction

The n-type semiconductor is formed by adding pentavalent impurities to the intrinsic semiconductor while p-type semiconductor is formed by adding trivalent impurities to the intrinsic semiconductor.

Also, in n-type semiconductors electrons are the majority carriers while holes are the minority carriers. On the other hand, in p-type semiconductors holes are the majority carriers while electrons are the minority carriers. The p-type and n-type semiconductors are not used separately for practical purpose because the overall charge of p-type and n-type semiconductors is electrically neutral. However, when p-type and n-type semiconductor materials are joined they behave differently.

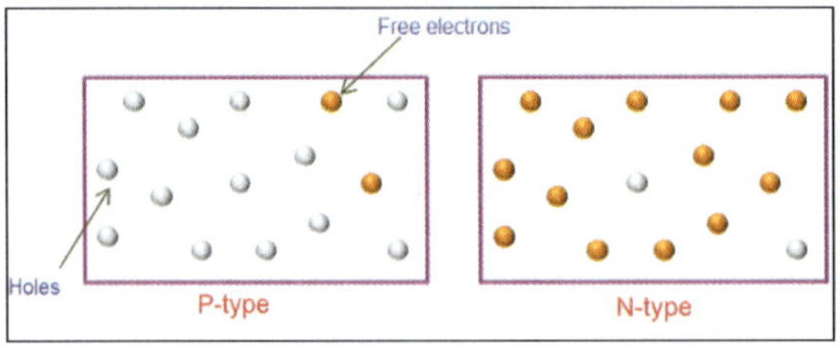

Remember that although n-type semiconductor has large number of free electrons, but these free electrons are provided by the pentavalent atoms that are electrically neutral. Thus, the n-type semiconductor is electrically neutral. Similarly, in p-type semiconductor holes are provided by the trivalent atoms that are electrically neutral. Therefore, overall charge of p-type semiconductor is also neutral.

Formation of p-n Junction

Generally junction refers to a point where two or more things are joined. For example, when one or more railway tracks are joined a railway junction is formed. The region where the tracks meet or joined is called railway junction.

In the similar way, when an n-type semiconductor is joined with the p-type semiconductor, a p-n junction is formed. The region where the p-type and n-type semiconductors are joined is called p-n junction. It is also defined as the boundary between p-type and n-type semiconductor. This p-n junction forms a most popular semiconductor device known as diode.

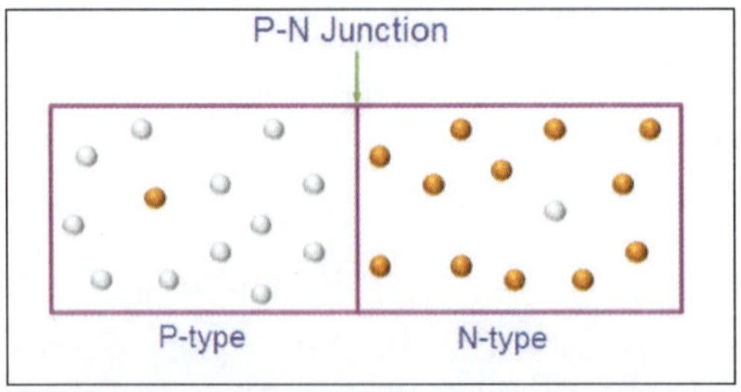

P-n junction is also a fundamental building block of many other semiconductor electronic devices such as transistors, solar cells, light emitting diodes, and integrated circuits. The credit of discovery of the p-n junction goes to American physicist Russel Ohi of Bell Laboratories.

Semiconductor devices are the fundamental building blocks of all the electronic devices such as computers, control systems, ATM (Automated Teller Machine), mobile phones, amplifiers, etc. Diodes are the simplest form of all the semiconductor devices. The various applications of diodes include computers, power supplies, television, radios and so on.

Before the invention of semiconductor diode there was vacuum tubes which are large in size, takes more power, costly, and noisy. This problem was solved with invention of semiconductor diode. Semiconductor diodes are small in size, low cost and consume less power.

Conduction Band in Semiconductors

The conduction band is the band of electron orbitals that electrons can jump up into from the valence band when excited. When the electrons are in these orbitals, they have enough energy to move freely in the material. This movement of electrons creates an electric current. The valence band is simply the outermost electron orbital of an atom of any specific material that electrons actually occupy. The energy difference between the highest occupied energy state of the valence band and the lowest unoccupied state of the conduction band is called the band gap and is indicative of the electrical conductivity of a material. A large band gap means that a lot of energy is required to excite valence electrons to the conduction band. Conversely, when the valence band and conduction band overlap as they do in metals, electrons can readily jump between the two bands meaning the material is highly conductive.

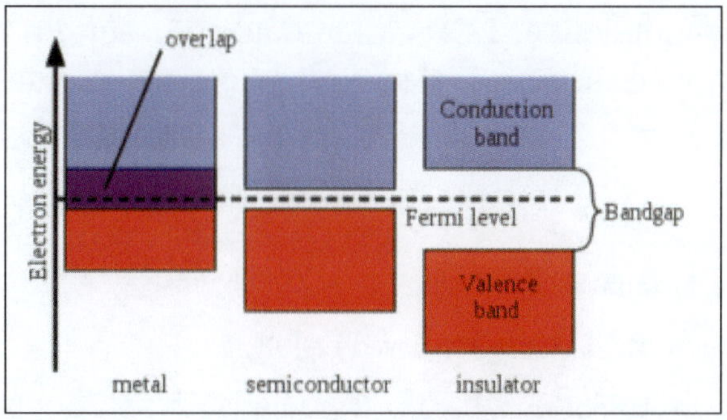

A diagram showing the valence and conduction bands of insulators, metals, and semiconductors. The Fermi level is the name given to the highest energy occupied electron orbital at absolute zero.

Insulators are characterized by a large band gap, so a prohibitively large amount of energy is required to move electrons into the conduction band to form a current. Conductors have an overlap between the conduction and valence bands, so the valence electrons in such conductors are essentially free. Semiconductors, on the other hand, have a small band gap that allows for a meaningful fraction of the valence electrons of the material to move into the conduction band given a certain amount of energy. This property gives them a conductivity between conductors and insulators, which is part of the reason why they are ideal for electric circuits as they will not cause a short circuit like a conductor. This band gap also allows semiconductors to convert light into electricity in photovoltaic cells and to emit light as LEDs when made into certain types of diodes. Both these processes rely on the energy absorbed or released by electrons moving between the conduction and valence bands.

Valence Band in Semiconductors

The valence band is the band of electron orbitals that electrons can jump out of, moving into the conduction band when excited. The valence band is simply the outermost electron orbital of an atom of any specific material that electrons actually occupy. This is closely related to the idea of the valence electron.

The energy difference between the highest occupied energy state of the valence band and the lowest unoccupied state of the conduction band is called the band gap and is indicative of the electrical conductivity of a material. A large band gap means that a lot of energy is required to excite valence electrons to the conduction band. Conversely, when the valence band and conduction band overlap as they do in metals, electrons can readily jump between the two bands meaning the material is highly conductive.

The difference between conductors, insulators, and semiconductors can be shown by how large their band gap is. Insulators are characterized by a large band gap, so a prohibitively large amount of energy is required to move electrons out of the valence band to form a current. Conductors have an overlap between the conduction and valence bands, so the valence electrons in such conductors are essentially free. Semiconductors, on the other hand, have a small band gap that allows for a meaningful fraction of the valence electrons of the material to move into the conduction band given a certain amount of energy. This property gives them a conductivity between conductors and insulators, which is part of the reason why they are ideal for circuits as they will not cause a short circuit like a conductor. This band gap also allows semiconductors to convert light into electricity in photovoltaic cells and to emit light as LEDs when made into certain types of diodes. Both these processes rely on the energy absorbed or released by electrons moving between the conduction and valence bands.

References

- Semiconductor, science: britannica.com, Retrieved 29 March, 2019
- N-type-semiconductor: circuitglobe.com, Retrieved 12 July, 2019
- P-type-semiconductor: circuitglobe.com, Retrieved 15 June ,2019
- Semiconductor-diodes/p-n-junction-introduction, electronic-devices-and-circuits: physics-and-radio-electronics.com, Retrieved 30 April, 2019
- Conduction-band, encyclopedia: energyeducation.ca, Retrieved 5 February, 2019
- Valence-band, encyclopedia: energyeducation.ca, Retrieved 3 May, 2019

5

Diodes: Types and Applications

A two-terminal electronic component that primarily conducts current in one direction is known as a diode. The main types of diodes are Zener diode, Schottky diode, tunnel diode, varactor diodes and Gunn diode. The diverse applications of these types of diodes have been thoroughly discussed in this chapter.

Diode is an electrical component that allows the flow of current in only one direction. In circuit diagrams, a diode is represented by a triangle with a line across one vertex.

The most common type of diode uses a p-n junction. In this type of diode, one material (n) in which electrons are charge carriers abuts a second material (p) in which holes (places depleted of electrons that act as positively charged particles) act as charge carriers. At their interface, a depletion region is formed across which electrons diffuse to fill holes in the p-side. This stops the further flow of electrons. When this junction is forward biased (that is, a positive voltage is applied to the p-side), electrons can easily move across the junction to fill the holes, and a current flows through the diode. When the junction is reverse biased (that is, a negative voltage is applied to the p-side), the depletion region widens and electrons cannot easily move across. The current remains very small until a certain voltage (the breakdown voltage) is reached and the current suddenly increases.

Light-emitting diodes (LEDs) are p-n junctions that emit light when a current flows through them. Several p-n junction diodes can be connected in series to make a rectifier (an electrical component that converts alternating current to direct current). Zener diodes have a well-defined breakdown voltage, so that current flows in the reverse direction at that voltage and a constant voltage can be maintained despite fluctuations in voltage or current. In varactor (or varicap) diodes, varying the bias voltage causes a variation in the diode's capacitance; these diodes have many applications for signal transmission and are used throughout the radio and television industries.

Early diodes were vacuum tubes, an evacuated glass or metal electron tube containing two electrodes—a negatively charged cathode and a positively charged anode. These were used as rectifiers and as detectors in electronic circuits such as radio and television receivers. When a positive voltage is applied to the anode (or plate), electrons emitted from the heated cathode flow to the plate and return to the cathode through an external power supply. If a negative voltage is applied to the plate, electrons cannot escape from the cathode, and no plate current flows. Thus, a diode permits electrons to flow from cathode to plate but not from plate to cathode. If an alternating voltage is applied to the plate, current flows only during the time when the plate is positive. The alternating voltage is said to be rectified, or converted to direct current.

Working Principle of Diode

Inside the p-n Junction Diode

The n side will have large number of electrons and very few holes (due to thermal excitation) whereas the p side will have high concentration of holes and very few electrons. Due to this a process called diffusion takes place. In this process free electrons from the n side will diffuse (spread) into the p side and combine with holes present there, leaving a positive immobile (not moveable) ion in the n side. Hence few atoms on the p side are converted into negative ions. Similarly few atoms on the n-side will get converted to positive ions. Due to this large number of positive ions and negative ions will accumulate on the n-side and p-side respectively. This region so formed is called as depletion region. Due to the presence of these positive and negative ions a static electric field called as "barrier potential" is created across the p-n junction of the diode. It is called as "barrier potential" because it acts as a barrier and opposes the flow of positive and negative ions across the junction.

To make use of this p-n junction diode we have to apply an external DC voltage to it. Applying an external DC voltage to the diode is called as biasing. If the p-side (anode) is connected to the positive terminal of the supply and the n-side (cathode) to the negative terminal of the supply, the diode is said to be forward biased. In the same way if the n-side is connected to the positive terminal of the supply and the p-side to the negative terminal of the supply, the diode is said to be reversed biased. Most of the times a resistance has to be connected in series with it to limit the electric current flowing through the diode. This is the working principle of diode.

Forward Biased Diode

When the diode is forward biased, due to the negative terminal on the n-side, electrons from the n-side are pushed towards the p-region. Similarly due to positive voltage on the p-side of the diode, Holes from the p-region are pushed towards n-side. Due to this the electrons will start converting the positive ions in the p-region into neutral atoms and holes will start converting the negative ions in the n-region to neutral atoms. Hence width of the depletion region starts reducing due to reduction in the barrier potential. (As the atoms in the depletion region are getting converted into neutral atoms less and less charged ions remain in this region with increase in supply voltage, hence width decreases.) This keeps happening and at a certain point the depletion region collapses and there is no opposition to the flow of current. Hence large number of electrons and holes will cross the junction and make the electric current to flow from anode to cathode.

Hence, forward biased electrical resistance of diode is very small and hence there is a small voltage drop (Practical condition, ideally there should be 0 forward resistance) across it. Its value for silicon diode is about 0.7 V.

Thus the p-n junction diode will allow a electric current to pass through it only when it is forward biased.

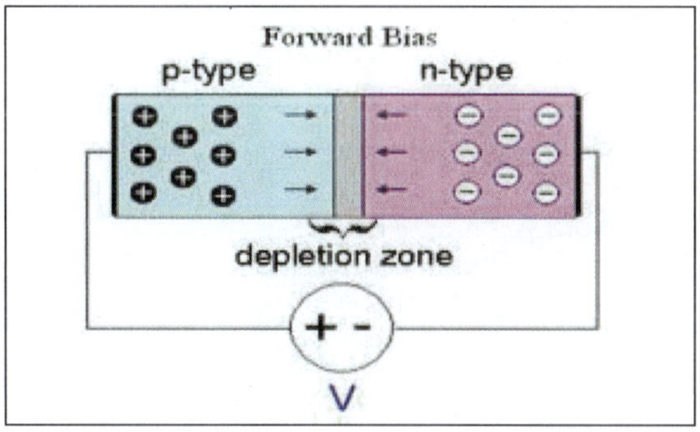

Reverse Biased Condition

When the diode is reverse biased the hole from the p-side will get attracted towards the negative terminal of the supply and electrons from the n-side are attracted towards the positive terminal. Hence the process of widening of the depletion region takes place and hence more and more opposition to the flow of electric current takes place.

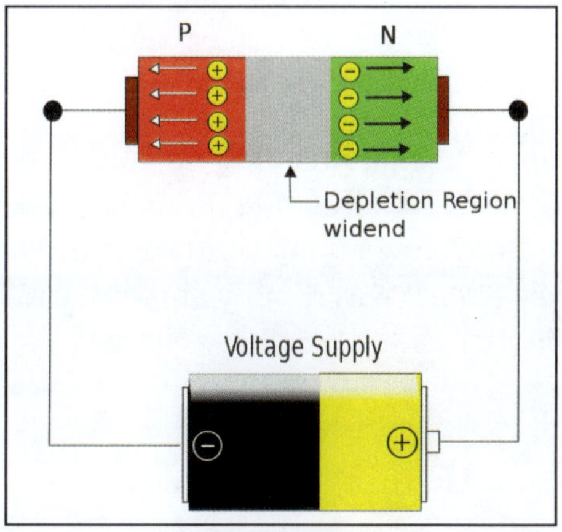

Hence, ideally the reverse biased resistance of the diode is infinite and no electric current flows from the diode when it is reversed biased. Due to large reverse biased voltage, suddenly large electric current will flow through the reverse biased voltage. Due to this large power gets dissipated in the diode which may damage it permanently.

V-I Characteristics of Diodes

The V-I characteristics or voltage-current characteristics of the p-n junction diode is shown in the figure. The horizontal line in the below figure represents the amount of voltage applied across the p-n junction diode whereas the vertical line represents the amount of current flows in the p-n junction diode.

- I_F = Forward Biased Current of diode;
- V_F = Forward Biased Voltage of diode;

- I_o = Reverse biased Saturation Current of Diode;
- I_R = Reverse biased Current of Diode.

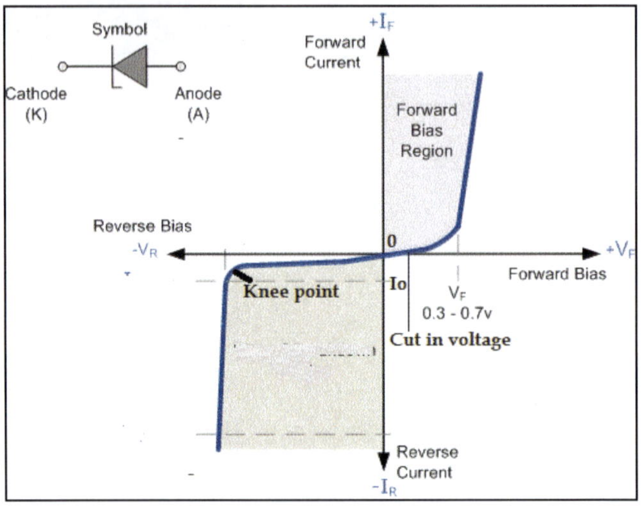

Forward Bias V-I Characteristic of P-N Diode

When anode is positive with respect to cathode, diode is said to be forward biased. With increase of the source voltage Vs from zero value, initially diode current is zero. From Vs=0 to cut-in voltage, the forward current is very small. Cut-in voltage is also known as threshold voltage or turn-on voltage. Beyond cut-in voltage, the diode current rises rapidly and diode said to conduct. For silicon diode, the cut-in voltage is around 0.7. When diode conducts, there is a forward voltage drop of the order of 0.8 to 1V.

Reverse Bias V-I Characteristic of P-N Diode

When cathode is positive with respect to anode the, the diode said to be reverse biased. In the reverse biased condition. A small reverse current leakage current, of the order of microampers or milliampers flow. The leakage current is almost independent of the reverse voltage until this voltage reach breakdown voltage at this reverse breakdown, voltage remains almost constant but reverse current becomes quite high limited only by the external circuit resistance. A large reverse break down voltage associated with high reverse current, leads to excessive power loss that may be destroy the diode.

At point a reverse breakdown of the diode occurs and current increase sharply damaging the diode. This point is called knee of the reverse characteristics.

V-I Characteristics of Typical Ge and Si Diode

- The cut in voltage voltage for germanium (Ge) diode is about 0.3 while for silicon (Si) diode is as about 0.7 V.
- The potential at which current starts increasing exponentially is called offset potential, threshold potential or firing potential of a diode.

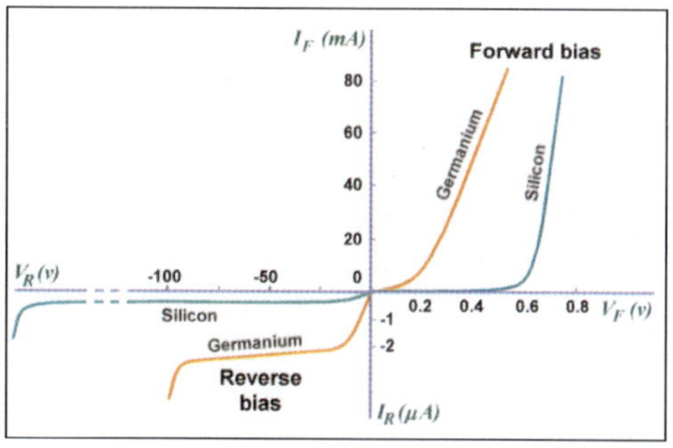

- The reverse saturation current is of order nA for silicon diode while it is of the order of uA for germanium diode.
- Reverse breakdown voltage for Si diode is higher than that of Ge diode of a comparable rating.

If the external reverse voltage applied on the p-n junction diode is increased, the free electrons from the n-type semiconductor and the holes from the p-type semiconductor are moved away from the p-n junction. This increases the width of depletion region.

The wide depletion region of reverse biased p-n junction diode completely blocks the majority charge carrier current. However, it allows the minority charge carrier current. The free electrons (minority carriers) in the p-type semiconductor and the holes (minority carriers) in the n-type semiconductor carry the electric current. The electric current, which is carried by the minority charge carriers in the p-n junction diode, is called reverse current.

In n-type and p-type semiconductors, very small number of minority charge carriers is present. Hence, a small voltage applied on the diode pushes all the minority carriers towards the junction. Thus, further increase in the external voltage does not increase the electric current. This electric current is called reverse saturation current. In other words, the voltage or point at which the electric current reaches its maximum level and further increase in voltage does not increase the electric current is called reverse saturation current.

The reverse saturation current is depends on the temperature. If temperature increases the generation of minority charge carriers increases. Hence, the reverse current increases with the increase in temperature. However, the reverse saturation current is independent of the external reverse voltage. Hence, the reverse saturation current remains constant with the increase in voltage. However, if the voltage applied on the diode is increased continuously, the p-n junction diode reaches to a state where junction breakdown occurs and reverse current increases rapidly.

In germanium diodes, a small increase in temperature generates large number of minority charge carriers. The number of minority charge carriers generated in the germanium diodes is greater than the silicon diodes. Hence, the reverse saturation current in the germanium diodes is greater than the silicon diodes.

Zener Diode

A Zener diode is a type of diode that allows current to flow not only from its anode to its cathode, but also in the reverse direction, when the *Zener voltage* is reached.

Zener diodes have a highly doped p–n junction. Normal diodes break down with a reverse voltage, but the voltage and sharpness of the knee are not as well defined as for a Zener diode. Normal diodes are not designed to operate in the breakdown region, whereas Zener diodes operate reliably in this region.

Zener reverse breakdown is due to electron quantum tunnelling caused by a high-strength electric field. However, many diodes described as "Zener" diodes rely instead on avalanche breakdown. Both breakdown types are used in Zener diodes with the Zener effect predominating at lower voltages and avalanche breakdown at higher voltages.

Zener diodes are widely used in electronic equipment of all kinds and are one of the basic building blocks of electronic circuits. They are used to generate low-power stabilized supply rails from a higher voltage and to provide reference voltages for circuits, especially stabilized power supplies. They are also used to protect circuits from overvoltage, especially electrostatic discharge (ESD).

Operation

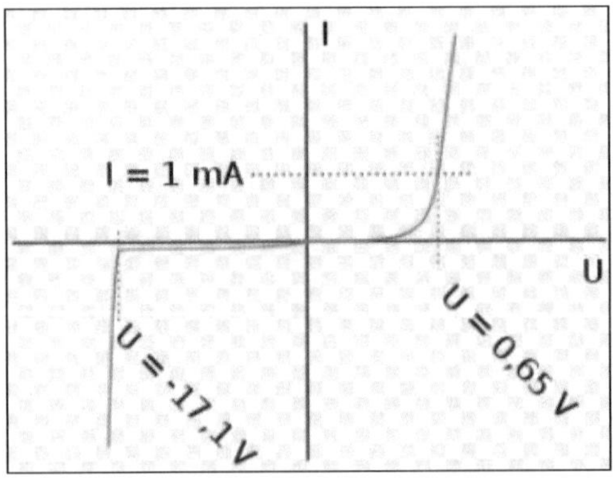

Current-voltage characteristic of a Zener diode with a breakdown voltage of 17 V. Notice the change of voltage scale between the forward biased (positive) direction and the reverse biased (negative) direction.

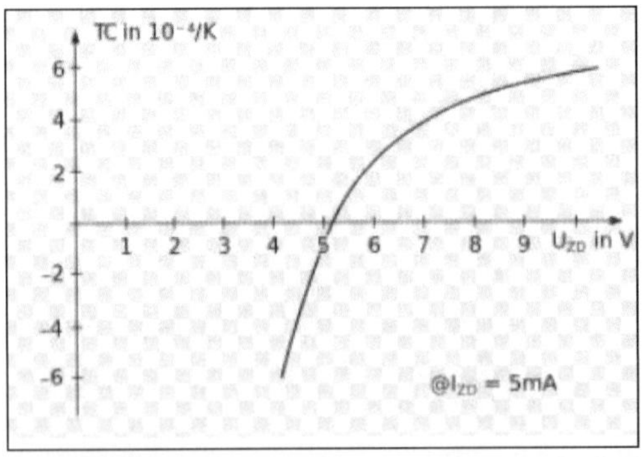

Temperature coefficient of Zener voltage against nominal Zener voltage.

A conventional solid-state diode allows significant current if it is reverse-biased above its reverse breakdown voltage. When the reverse bias breakdown voltage is exceeded, a conventional diode is subject to high current due to avalanche breakdown. Unless this current is limited by circuitry, the diode may be permanently damaged due to overheating. A Zener diode exhibits almost the same properties, except the device is specially designed so as to have a reduced breakdown voltage, the so-called Zener voltage. By contrast with the conventional device, a reverse-biased Zener diode exhibits a controlled breakdown and allows the current to keep the voltage across the Zener diode close to the Zener breakdown voltage. For example, a diode with a Zener breakdown voltage of 3.2 V exhibits a voltage drop of very nearly 3.2 V across a wide range of reverse currents. The Zener diode is therefore ideal for applications such as the generation of a reference voltage (e.g. for an amplifier stage), or as a voltage stabilizer for low-current applications.

Another mechanism that produces a similar effect is the avalanche effect as in the avalanche diode. The two types of diode are in fact constructed the same way and both effects are present in diodes of this type. In silicon diodes up to about 5.6 volts, the Zener effect is the predominant effect and shows a marked negative temperature coefficient. Above 5.6 volts, the avalanche effect becomes predominant and exhibits a positive temperature coefficient.

In a 5.6 V diode, the two effects occur together, and their temperature coefficients nearly cancel each other out, thus the 5.6 V diode is useful in temperature-critical applications. An alternative, which is used for voltage references that need to be highly stable over long periods of time, is to use a Zener diode with a temperature coefficient (TC) of +2 mV/°C (breakdown voltage 6.2–6.3 V) connected in series with a forward-biased silicon diode (or a transistor B-E junction) manufactured on the same chip. The forward-biased diode has a temperature coefficient of −2 mV/°C, causing the TCs to cancel out.

Modern manufacturing techniques have produced devices with voltages lower than 5.6 V with negligible temperature coefficients, but as higher-voltage devices are encountered, the temperature coefficient rises dramatically. A 75 V diode has 10 times the coefficient of a 12 V diode.

Zener and avalanche diodes, regardless of breakdown voltage, are usually marketed under the umbrella term of "Zener diode".

Under 5.6 V, where the Zener effect dominates, the IV curve near breakdown is much more rounded, which calls for more care in targeting its biasing conditions. The IV curve for Zeners above 5.6 V (being dominated by Avalanche), is much sharper at breakdown.

Construction

The Zener diode's operation depends on the heavy doping of its p-n junction. The depletion region formed in the diode is very thin (<1 µm) and the electric field is consequently very high (about 500 kV/m) even for a small reverse bias voltage of about 5 V, allowing electrons to tunnel from the valence band of the p-type material to the conduction band of the n-type material.

At the atomic scale, this tunneling corresponds to the transport of valence band electrons into the empty conduction band states; as a result of the reduced barrier between these bands and high electric fields that are induced due to the high levels of doping on both sides. The breakdown voltage can be controlled quite accurately in the doping process. While tolerances within 0.07% are available, the most widely used tolerances are 5% and 10%. Breakdown voltage for commonly available Zener diodes can vary widely from 1.2 V to 200 V.

For diodes that are lightly doped the breakdown is dominated by the avalanche effect rather than the Zener effect. Consequently, the breakdown voltage is higher (over 5.6 V) for these devices.

Surface Zeners

The emitter-base junction of a bipolar NPN transistor behaves as a Zener diode, with breakdown voltage at about 6.8 V for common bipolar processes and about 10 V for lightly doped base regions in BiCMOS processes. Older processes with poor control of doping characteristics had the variation of Zener voltage up to ±1 V, newer processes using ion implantation can achieve no more than ±0.25 V. The NPN transistor structure can be employed as a *surface Zener diode*, with collector and emitter connected together as its cathode and base region as anode. In this approach the base doping profile usually narrows towards the surface, creating a region with intensified electric field where the avalanche breakdown occurs. The hot carriers produced by acceleration in the intense field sometime shoot into the oxide layer above the junction and become trapped there. The accumulation of trapped charges can then cause 'Zener walkout', a corresponding change of the Zener voltage of the junction. The same effect can be achieved by radiation damage.

The emitter-base Zener diodes can handle only smaller currents as the energy is dissipated in the base depletion region which is very small. Higher amount of dissipated energy (higher current for longer time, or a short very high current spike) causes thermal damage to the junction and/or its contacts. Partial damage of the junction can shift its Zener voltage. Total destruction of the Zener junction by overheating it and causing migration of metallization across the junction ("spiking") can be used intentionally as a 'Zener zap' antifuse.

Subsurface Zeners

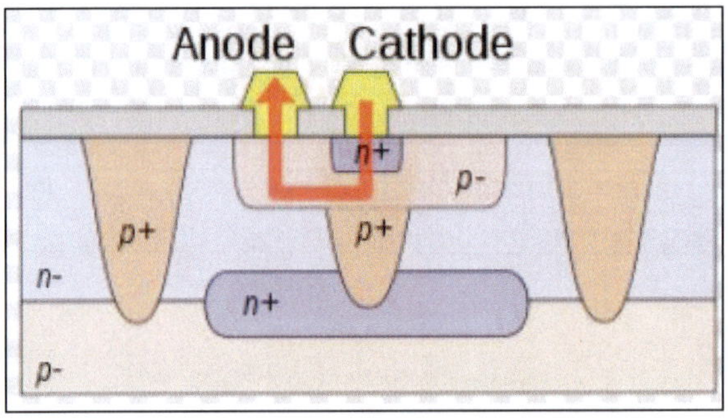

Buried zener structure.

A subsurface Zener diode, also called 'buried Zener', is a device similar to the Surface Zener, but with the avalanche region located deeper in the structure, typically several micrometers below the oxide. The hot carriers then lose energy by collisions with the semiconductor lattice before reaching the oxide layer and cannot be trapped there. The Zener walkout phenomenon therefore does not occur here, and the buried Zeners have voltage constant over their entire lifetime. Most buried Zeners have breakdown voltage of 5–7 volts. Several different junction structures are used.

Uses

Zener diodes are widely used as voltage references and as shunt regulators to regulate the voltage across small circuits. When connected in parallel with a variable voltage source so that it is reverse biased, a Zener diode conducts when the voltage reaches the diode's reverse breakdown voltage. From that point on, the low impedance of the diode keeps the voltage across the diode at that value.

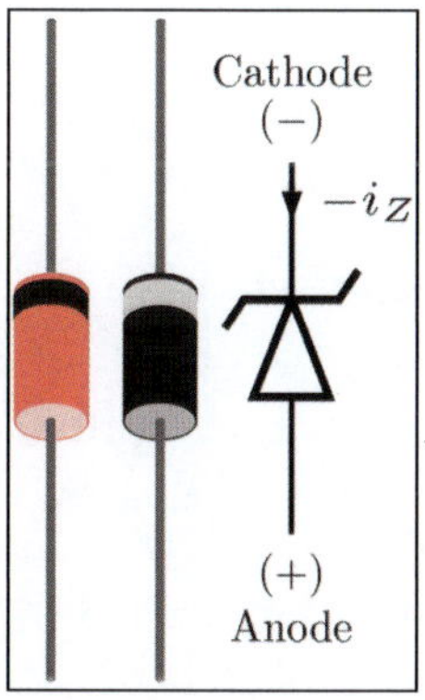

Zener diode shown with typical packages. *Reverse* current $-i_Z$ is shown.

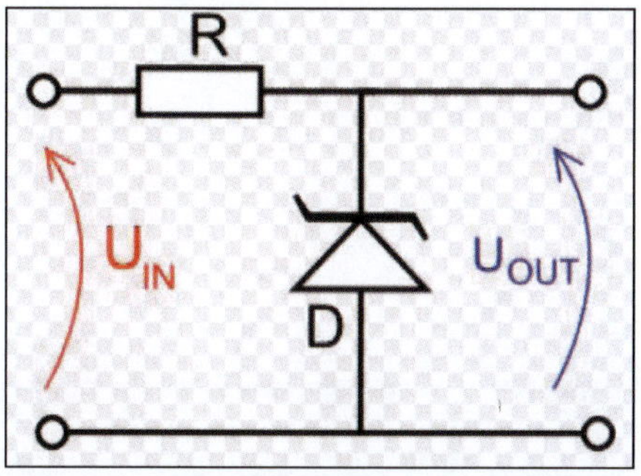

In this circuit, a typical voltage reference or regulator, an input voltage, U_{in}, is regulated down to a stable output voltage U_{out}. The breakdown voltage of diode D is stable over a wide current range and holds U_{out} approximately constant even though the input voltage may fluctuate over a wide range. Because of the low impedance of the diode when operated like this, resistor *R* is used to limit current through the circuit.

In the case of this simple reference, the current flowing in the diode is determined using Ohm's law and the known voltage drop across the resistor *R*;

$$I_{diode} = \frac{U_{in} - U_{out}}{R}$$

The value of *R* must satisfy two conditions:

1. *R* must be small enough that the current through D keeps D in reverse breakdown. The value of this current is given in the data sheet for D. For example, the common BZX79C5V6 device, a 5.6 V 0.5 W Zener diode, has a recommended reverse current of 5 mA. If insufficient current exists through D, then U_{out} is unregulated and less than the nominal breakdown voltage (this differs from voltage-regulator tubes where the output voltage is higher than nominal and could rise as high as U_{in}). When calculating *R*, allowance must be made for any current through the external load, not shown in this diagram, connected across U_{out}.
2. *R* must be large enough that the current through D does not destroy the device. If the current through D is I_D, its breakdown voltage V_B and its maximum power dissipation P_{max} correlate as such: $I_D V_B < P_{max}$.

A load may be placed across the diode in this reference circuit, and as long as the Zener stays in reverse breakdown, the diode provides a stable voltage source to the load. Zener diodes in this configuration are often used as stable references for more advanced voltage regulator circuits.

Shunt regulators are simple, but the requirements that the ballast resistor be small enough to avoid excessive voltage drop during worst-case operation (low input voltage concurrent with high load current) tends to leave a lot of current flowing in the diode much of the time, making for a fairly wasteful regulator with high quiescent power dissipation, only suitable for smaller loads.

These devices are also encountered, typically in series with a base-emitter junction, in transistor stages where selective choice of a device centered on the avalanche or Zener point can be used to introduce compensating temperature co-efficient balancing of the transistor p–n junction. An example of this kind of use would be a DC error amplifier used in a regulated power supply circuit feedback loop system.

Zener diodes are also used in surge protectors to limit transient voltage spikes.

Another application of the Zener diode is the use of noise caused by its avalanche breakdown in a random number generator.

Waveform Clipper

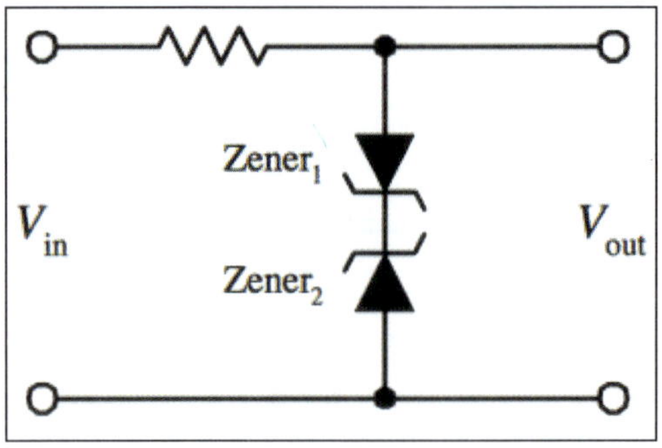

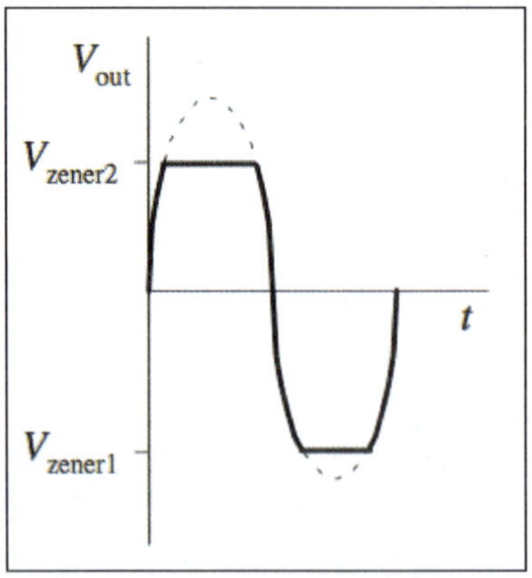

Examples of a waveform clipper.

Two Zener diodes facing each other in series clip both halves of an input signal. Waveform clippers can be used not only to reshape a signal, but also to prevent voltage spikes from affecting circuits that are connected to the power supply.

Voltage Shifter

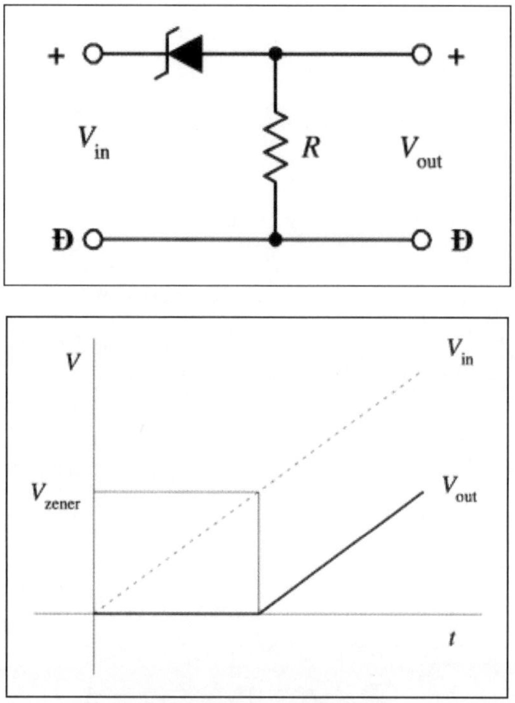

Examples of a voltage shifter.

A Zener diode can be applied to a circuit with a resistor to act as a voltage shifter. This circuit lowers the output voltage by a quantity that is equal to the Zener diode's breakdown voltage.

Voltage Regulator

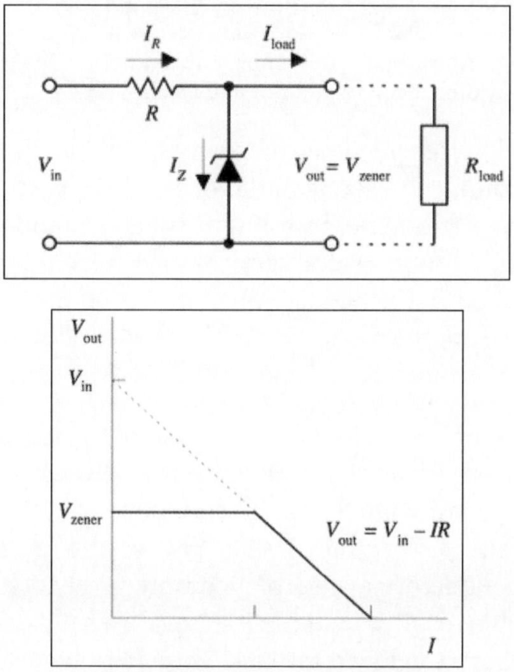

Examples of a voltage regulator

A Zener diode can be applied in a voltage regulator circuit to regulate the voltage applied to a load, such as in a linear regulator.

Schottky Diode

The Schottky diode (named after the German physicist Walter H. Schottky), also known as Schottky barrier diode or hot-carrier diode, is a semiconductor diode formed by the junction of a semiconductor with a metal. It has a low forward voltage drop and a very fast switching action. The cat's-whisker detectors used in the early days of wireless and metal rectifiers used in early power applications can be considered primitive Schottky diodes.

When sufficient forward voltage is applied, a current flows in the forward direction. A silicon diode has a typical forward voltage of 600–700 mV, while the Schottky's forward voltage is 150–450 mV. This lower forward voltage requirement allows higher switching speeds and better system efficiency.

Construction

1N5822 Schottky diode with cut-open packaging. The semiconductor in the center makes a Schottky barrier against one metal electrode (providing rectifying action) and an ohmic contact with the other electrode.

A metal–semiconductor junction is formed between a metal and a semiconductor, creating a Schottky barrier (instead of a semiconductor–semiconductor junction as in conventional diodes). Typical metals used are molybdenum, platinum, chromium or tungsten, and certain silicides (e.g., palladium silicide and platinum silicide), whereas the semiconductor would typically be n-type silicon. The metal side acts as the anode, and n-type semiconductor acts as the cathode of the diode; meaning conventional current can flow from the metal side to the semiconductor side, but not in the opposite direction. This Schottky barrier results in both very fast switching and low forward voltage drop.

The choice of the combination of the metal and semiconductor determines the forward voltage of the diode. Both n- and p-type semiconductors can develop Schottky barriers. However, the p-type typically has a much lower forward voltage. As the reverse leakage current increases dramatically with lowering the forward voltage, it cannot be too low, so the usually employed range is about 0.5–0.7 V, and p-type semiconductors are employed only rarely. Titanium silicide and other refractory silicides, which are able to withstand the temperatures needed for source/drain annealing in CMOS processes, usually have too low a forward voltage to be useful, so processes using these silicides therefore usually do not offer Schottky diodes.

With increased doping of the semiconductor, the width of the depletion region drops. Below a certain width, the charge carriers can tunnel through the depletion region. At very high doping levels, the junction does not behave as a rectifier any more and becomes an ohmic contact. This can be used for the simultaneous formation of ohmic contacts and diodes, as a diode will form between the silicide and lightly doped n-type region, and an ohmic contact will form between the silicide and the heavily doped n- or p-type region. Lightly doped p-type regions pose a problem, as the resulting contact has too high a resistance for a good ohmic contact, but too low a forward voltage and too high a reverse leakage to make a good diode.

As the edges of the Schottky contact are fairly sharp, a high electric field gradient occurs around them, which limits how large the reverse breakdown voltage threshold can be. Various strategies are used, from guard rings to overlaps of metallization to spread out the field gradient. The guard rings consume valuable die area and are used primarily for larger higher-voltage diodes, while overlapping metallization is employed primarily with smaller low-voltage diodes.

Schottky diodes are often used as antisaturation clamps in Schottky transistors. Schottky diodes made from palladium silicide (PdSi) are excellent due to their lower forward voltage (which has to be lower than the forward voltage of the base-collector junction). The Schottky temperature coefficient is lower than the coefficient of the B–C junction, which limits the use of PdSi at higher temperatures.

For power Schottky diodes, the parasitic resistances of the buried n+ layer and the epitaxial n-type layer become important. The resistance of the epitaxial layer is more important than it is for a transistor, as the current must cross its entire thickness. However, it serves as a distributed ballasting resistor over the entire area of the junction and, under usual conditions, prevents localized thermal runaway.

In comparison with the power p–n diodes the Schottky diodes are less rugged. The junction is direct contact with the thermally sensitive metallization, a Schottky diode can therefore dissipate less power than an equivalent-size p-n counterpart with a deep-buried junction before failing (especially during reverse breakdown). The relative advantage of the lower forward voltage of Schottky diodes is diminished at higher forward currents, where the voltage drop is dominated by the series resistance.

Reverse Recovery Time

The most important difference between the p-n diode and the Schottky diode is the reverse recovery time (t_{rr}), when the diode switches from the conducting to the non-conducting state. In a p–n diode, the reverse recovery time can be in the order of several microseconds to less than 100 ns for fast diodes. Schottky diodes do not have a recovery time, as there is nothing to recover from (i.e., there is no charge carrier depletion region at the junction). The switching time is ~100 ps for the small-signal diodes, and up to tens of nanoseconds for special high-capacity power diodes. With p–n-junction switching, there is also a reverse recovery current, which in high-power semiconductors brings increased EMI noise. With Schottky diodes, switching is essentially "instantaneous" with only a slight capacitive loading, which is much less of a concern.

This "instantaneous" switching is not always the case. In higher voltage Schottky devices, in particular, the guard ring structure needed to control breakdown field geometry creates a parasitic p-n diode with the usual recovery time attributes. As long as this guard ring diode is not forward biased, it adds only capacitance. If the Schottky junction is driven hard enough however, the forward voltage eventually will bias both diodes forward and actual t_{rr} will be greatly impacted.

It is often said that the Schottky diode is a "majority carrier" semiconductor device. This means that if the semiconductor body is a doped n-type, only the n-type carriers (mobile electrons) play a significant role in normal operation of the device. The majority carriers are quickly injected into the conduction band of the metal contact on the other side of the diode to become free moving electrons. Therefore, no slow random recombination of n and p type carriers is involved, so that this diode can cease conduction faster than an ordinary p–n rectifier diode. This property in turn allows a smaller device area, which also makes for a faster transition. This is another reason why Schottky diodes are useful in switch-mode power converters: the high speed of the diode means that the circuit can operate at frequencies in the range 200 kHz to 2 MHz, allowing the use of small inductors and capacitors with greater efficiency than would be possible with other diode types. Small-area Schottky diodes are the heart of RF detectors and mixers, which often operate at frequencies up to 50 GHz.

Limitations

The most evident limitations of Schottky diodes are their relatively low reverse voltage ratings, and their relatively high reverse leakage current. For silicon-metal Schottky diodes, the reverse voltage is typically 50 V or less. Some higher-voltage designs are available (200 V is considered a high reverse voltage). Reverse leakage current, since it increases with temperature, leads to a thermal instability issue. This often limits the useful reverse voltage to well below the actual rating.

While higher reverse voltages are achievable, they would present a higher forward voltage, comparable to other types of standard diodes. Such Schottky diodes would have no advantage unless great switching speed is required.

Silicon Carbide Schottky Diode

Schottky diodes constructed from silicon carbide have a much lower reverse leakage current than silicon Schottky diodes, as well as higher forward voltage (about 1.4-1.8V at 25 °C) and reverse voltage. As of 2011 they were available from manufacturers in variants up to 1700 V of reverse voltage.

Silicon carbide has a high thermal conductivity, and temperature has little influence on its switching and thermal characteristics. With special packaging, silicon carbide Schottky diodes can operate at junction temperatures of over 500 K (about 200 °C), which allows passive radiative cooling in aerospace applications.

Applications

Voltage Clamping

While standard silicon diodes have a forward voltage drop of about 0.6 V and germanium diodes 0.2 V, Schottky diodes' voltage drop at forward biases of around 1 mA is in the range of 0.15 V to 0.46 V, which makes them useful in voltage clamping applications and prevention of transistor saturation. This is due to the higher current density in the Schottky diode.

Reverse Current and Discharge Protection

Because of a Schottky diode's low forward voltage drop; less energy is wasted as heat, making them the most efficient choice for applications sensitive to efficiency. For instance, they are used in stand-alone ("off-grid") photovoltaic (PV) systems to prevent batteries from discharging through the solar panels at night, called "blocking diodes". They are also used in grid-connected systems with multiple strings connected in parallel, in order to prevent reverse current flowing from adjacent strings through shaded strings if the "bypass diodes" have failed.

Switched-mode Power Supplies

Schottky diodes are also used as rectifiers in switched-mode power supplies. The low forward voltage and fast recovery time leads to increased efficiency.

They can also be used in power supply "OR"ing circuits in products that have both an internal battery and a mains adapter input, or similar. However, the high reverse leakage current presents a problem in this case, as any high-impedance voltage sensing circuit (e.g., monitoring the battery voltage or detecting whether a mains adapter is present) will see the voltage from the other power source through the diode leakage.

Sample-and-hold Circuits

Schottky diodes can be used in diode-bridge based sample and hold circuits. When compared to regular p-n junction based diode bridges, Schottky diodes can offer advantages. A forward-biased Schottky diode does not have any minority carrier charge storage. This allows them to switch more quickly than regular diodes, resulting in lower transition time from the sample to the hold step. The absence of minority carrier charge storage also results in a lower hold step or sampling error, resulting in a more accurate sample at the output.

Charge Control

Due to its efficient electric field control Schottky diodes can be used to accurately load or unload single electrons in semiconductor nanostructures such as quantum wells or quantum dots.

Designation

SS14 schottky diode in a DO-214AC (SMA) package (surface mount version of 1N5819).

Commonly encountered schottky diodes include the 1N58xx series rectifiers, such as the 1N581x (1 ampere) and 1N582x (3 ampere) through-hole parts, and the SS1x (1 ampere) and SS3x (3 ampere) surface-mount parts. Schottky rectifiers are available in numerous surface-mount package styles.

Small-signal schottky diodes such as the 1N5711, 1N6263, 1SS106, 1SS108, and the BAT41–43, 45–49 series are widely used in high-frequency applications as detectors, mixers and nonlinear elements, and have superseded germanium diodes. They are also suitable for electrostatic discharge (ESD) protection of sensitive devices such as III-V-semiconductor devices, laser diodes and, to a lesser extent, exposed lines of CMOS circuitry.

Schottky metal–semiconductor junctions are featured in the successors to the 7400 TTL family of logic devices, the 74S, 74LS and 74ALS series, where they are employed as Baker clamps in parallel with the collector-base junctions of the bipolar transistors to prevent their saturation, thereby greatly reducing their turn-off delays.

Alternatives

When less power dissipation is desired, a MOSFET and a control circuit can be used instead, in an operation mode known as active rectification.

A super diode consisting of a pn-diode or Schottky diode and an operational amplifier provides an almost perfect diode characteristic due to the effect of negative feedback, although its use is restricted to frequencies the operational amplifier used can handle.

Electrowetting

Electrowetting can be observed when a Schottky diode is formed using a droplet of liquid metal, e.g. mercury, in contact with a semiconductor, e.g. silicon. Depending on the doping type and density in the semiconductor, the droplet spreading depends on the magnitude and sign of the voltage applied to the mercury droplet. This effect has been termed 'Schottky electrowetting'.

Tunnel Diode

A tunnel diode or Esaki diode is a type of semiconductor diode that has negative resistance due to the quantum mechanical effect called tunneling. It was invented in August 1957 by Leo Esaki, Yuriko Kurose, and Takashi Suzuki when they were working at Tokyo Tsushin Kogyo, now known as Sony. In 1973, Esaki received the Nobel Prize in Physics, jointly with Brian Josephson, for discovering the electron tunneling effect used in these diodes. Robert Noyce independently devised the idea of a tunnel diode while working for William Shockley, but was discouraged from pursuing it. Tunnel diodes were first manufactured by Sony in 1957, followed by General Electric and other companies from about 1960, and are still made in low volume today.

Tunnel diodes have a heavily doped p–n junction that is about 10 nm (100 Å) wide. The heavy doping results in a broken band gap, where conduction band electron states on the n-side are more or less aligned with valence band hole states on the p-side. They are usually made from germanium, but can also be made from gallium arsenide and silicon materials. Their negative differential resistance in part of their operating range allows them to function as oscillators and amplifiers, and in switching circuits using hysteresis. They are also used as frequency converters and detectors. Their low capacitance allows them to function at microwave frequencies, above the range of ordinary diodes and transistors.

Tunnel diodes are not widely used due to their low output power; their RF output is limited to several hundred milliwatts due to their small voltage swing. In recent years, however, new devices that use the tunneling mechanism have been developed. The resonant-tunneling diode (RTD) has achieved some of the highest frequencies of any solid-state oscillator. Another type of tunnel diode is a metal–insulator–metal (MIM) diode, but its present application appears to be limited to research environments due to inherent sensitivities. There is also a metal–insulator–insulator–metal (MIIM) diode, where an additional insulator layer allows "*step tunneling*" for precise diode control.

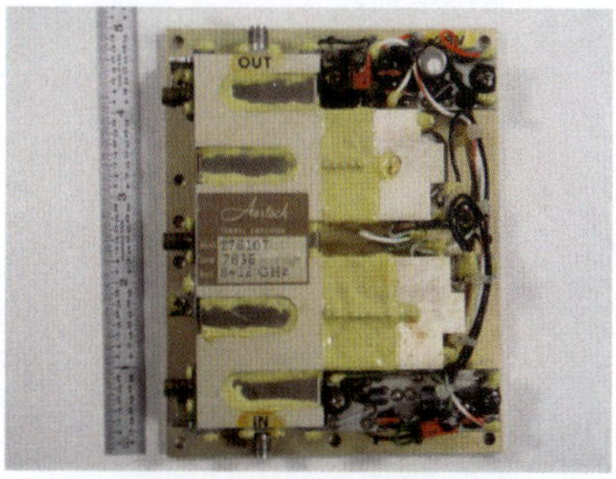

8–12 GHz tunnel diode amplifier, circa 1970.

Forward Bias Operation

Under normal forward bias operation, as voltage begins to increase, electrons at first tunnel through the very narrow p–n junction barrier and fill electron states in the conduction band on the n-side which become aligned with empty valence band hole states on the p-side of the p-n junction. As voltage increases further, these states become increasingly misaligned, and the current drops. This is called negative differential resistance because current decreases with increasing voltage. As voltage increases, the diode begins to operate as a normal diode, where electrons travel by conduction across the p–n junction, and no longer by tunneling through the p–n junction barrier. The most important operating region for a tunnel diode is the negative resistance region. Its graph is different from normal p–n junction diode.

Reverse Bias Operation

When used in the reverse direction, tunnel diodes are called back diodes (or backward diodes) and can act as fast rectifiers with zero offset voltage and extreme linearity for power signals (they have an accurate square law characteristic in the reverse direction). Under reverse bias, filled states on the p-side become increasingly aligned with empty states on the n-side, and electrons now tunnel through the p–n junction barrier in reverse direction.

Technical Comparisons

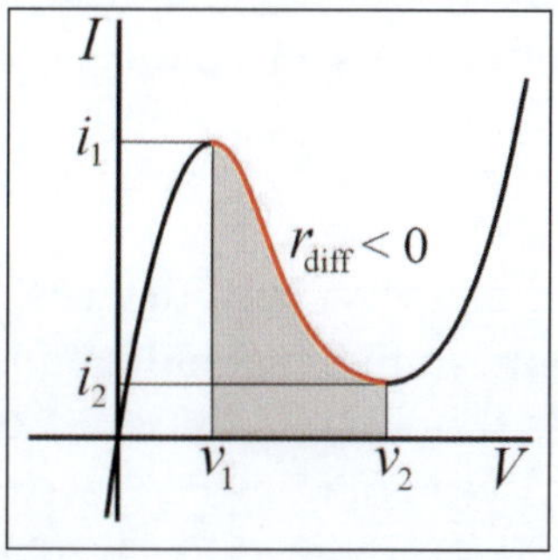

IV curve similar to a tunnel diode characteristic curve. It has negative differential resistance in the shaded voltage region, between v_1 and v_2.

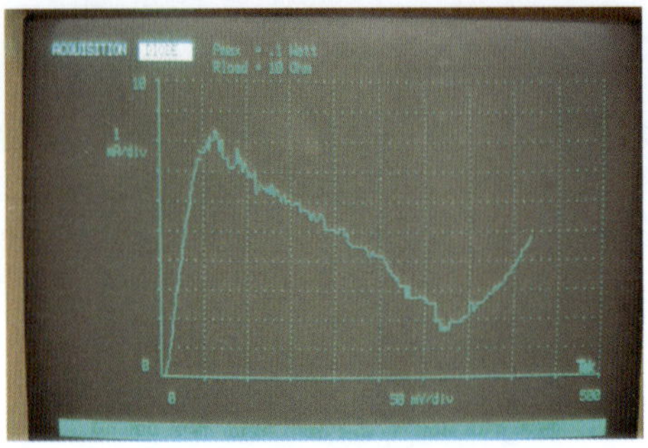

I-V curve of 10 mA germanium tunnel diode, taken on a Tektronix model 571 curve tracer.

In a conventional semiconductor diode, conduction takes place while the p–n junction is forward biased and blocks current flow when the junction is reverse biased. This occurs up to a point known as the "reverse breakdown voltage" at which point conduction begins (often accompanied by destruction of the device). In the tunnel diode, the dopant concentrations in the p and n layers are increased to a level such that the reverse breakdown voltage becomes zero and the diode conducts in the reverse direction. However, when forward-biased, an effect occurs called quantum mechanical tunneling which gives rise to a region in its voltage-current behavior where an *increase* in forward voltage is accompanied by a *decrease* in forward current. This negative resistance region can be exploited in a solid state version of the dynatron oscillator which normally uses a tetrode thermionic valve (vacuum tube).

Applications

The tunnel diode showed great promise as an oscillator and high-frequency threshold (trigger) device since it operated at frequencies far greater than the tetrode could: well into the microwave bands. Applications of tunnel diodes included local oscillators for UHF television tuners, trigger circuits in oscilloscopes, high-speed counter circuits, and very fast-rise time pulse generator circuits. In 1977, the Intelsat V satellite receiver used a microstrip tunnel diode amplifier (TDA) front-end in the 14 to 15.5 GHz frequency band. Such amplifiers were considered state-of-the-art, with better performance at high frequencies than any transistor-based front end. The tunnel diode can also be used as a low-noise microwave amplifier. Since its discovery, more conventional semiconductor devices have surpassed its performance using conventional oscillator techniques. For many purposes, a three-terminal device, such as a field-effect transistor, is more flexible than a device with only two terminals. Practical tunnel diodes operate at a few milliamperes and a few tenths of a volt, making them low-power devices. The Gunn diode has similar high frequency capability and can handle more power.

Tunnel diodes are also more resistant to ionizing radiation than other diodes. This makes them well suited to higher radiation environments such as those found in space.

Longevity

Tunnel diodes are notable for their longevity, with devices made in the 1960s still functioning. Writing in *Nature*, Esaki and coauthors state that semiconductor devices in general are extremely stable, and suggest that their shelf life should be "infinite" if kept at room temperature. They go on to report that a small-scale test of 50-year-old devices revealed a "gratifying confirmation of the diode's longevity". As noticed on some samples of Esaki diodes, the gold plated iron pins can in fact corrode and short out to the case. This can usually be diagnosed and treated with simple peroxide/vinegar technique normally used for repairing phone PCBs and the diode inside normally still works.

These components are susceptible to damage by overheating, and thus special care is needed when soldering them. Surplus Russian units are also reliable and often can be purchased for a few pence despite original cost being in the £30–50 range. The units typically sold are GaAs based and have a I_{pk}/I_v ratio of 5:1 at around 1–20 mA I_{pk}, and so should be protected against overcurrent.

Varactor Diodes

In electronics, a varicap diode, varactor diode, variable capacitance diode, variable reactance diode or tuning diode is a type of diode designed to exploit the voltage-dependent capacitance of a reverse-biased p–n junction.

Applications

Varactors are used as voltage-controlled capacitors. They are commonly used in voltage-controlled oscillators, parametric amplifiers, and frequency multipliers. Voltage-controlled oscillators have many applications such as frequency modulation for FM transmitters and phase-locked loops. Phase-locked loops are used for the frequency synthesizers that tune many radios, television sets, and cellular telephones.

The varicap was developed by the Pacific Semiconductor subsidiary of the Ramo Wooldridge Corporation who received a patent for the device in June 1961. The device name was also trademarked as the "Varicap" by TRW Semiconductors, the successor to Pacific Semiconductors, in October 1967. This helps explain the different names for the device as it came into use.

Operation

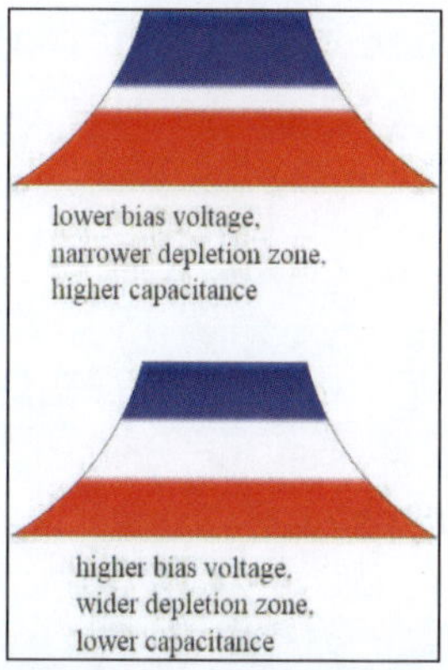

Operation of a varicap. Holes are blue, electrons are red, depletion zone is white. The electrodes are at the top and bottom.

Varactors are operated in a reverse-biased state, so no DC current flows through the device. The amount of reverse bias controls the thickness of the depletion zone and therefore the varactor's junction capacitance. Generally, the depletion region thickness is proportional to the square root of the applied voltage, and capacitance is inversely proportional to the depletion region thickness. Thus, the capacitance is inversely proportional to the square root of applied voltage.

All diodes exhibit this variable junction capacitance, but varactors are manufactured to exploit the effect and increase the capacitance variation.

The figure shows an example of a cross section of a varactor with the depletion layer formed of a p–n junction. This depletion layer can also be made of a MOS or a Schottky diode. This is important in CMOS and MMIC technology.

Use in a Circuit

Tuning Circuits

Generally the use of a varicap diode in a circuit requires connecting it to a tuned circuit, usually in parallel with any existing capacitance or inductance. A DC voltage is applied as reverse bias across the varicap to alter its capacitance. The DC bias voltage must be blocked from entering the tuned circuit. This can be accomplished by placing a DC blocking capacitor with a capacitance about 100 times greater than the maximum capacitance of the varicap diode in series with it and by applying DC from a high impedance source to the node between the varicap cathode and the blocking capacitor as shown in the upper left circuit in the accompanying diagram.

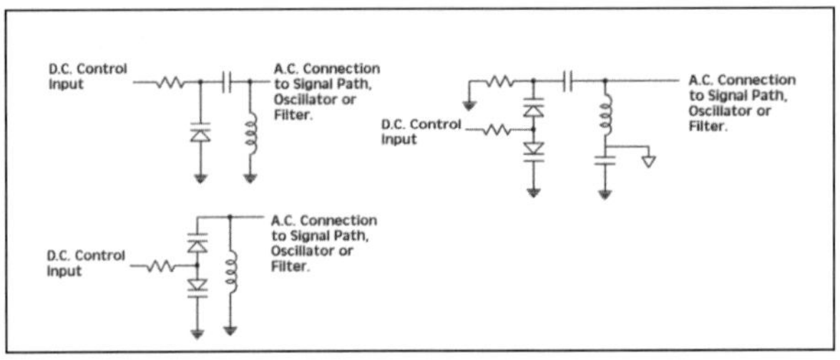

Since no significant DC current flows in the varicap, the value of the resistor connecting its cathode back to the DC control voltage resistor can be somewhere in the range of 22 kΩ to 150 kΩ and the blocking capacitor somewhere in the range of 5–100 nF. Sometimes, with very high-Q tuned circuits, an inductor is placed in series with the resistor to increase the source impedance of the control voltage so as not to load the tuned circuit and decrease its Q.

Another common configuration uses two back-to-back (anode to anode) varicap diodes. The second varicap effectively replaces the blocking capacitor in the first circuit. This reduces the overall capacitance and the capacitance range by half, but has the advantage of reducing the AC component of voltage across each device and has symmetrical distortion should the AC component possess enough amplitude to bias the varicaps into forward conduction.

When designing tuning circuits with varicaps it is usually good practice to maintain the AC component of voltage across the varicap at a minimal level, usually less than 100 mV peak to peak, to prevent changing the diode capacitance too much, which would distort the signal and add harmonics.

A third circuit, at top right in diagram, uses two series-connected varicaps and separate DC and AC signal ground connections. The DC ground is shown as a traditional ground symbol, and the AC ground as an open triangle. Separation of grounds is often done to (i) prevent high-frequency radiation from the low-frequency ground node, and (ii) prevent DC currents in the AC ground node changing bias and operating points of active devices such as varicaps and transistors.

These circuit configurations are quite common in television tuners and electronically tuned broadcast AM and FM receivers, as well as other communications equipment and industrial equipment. Early varicap diodes usually required a reverse voltage range of 0–33 V to obtain their full capacitance ranges, which were still quite small, approximately 1–10 pF. These types were – and still are – extensively used in television tuners, whose high carrier frequencies require only small changes in capacitance.

In time, varicap diodes were developed which exhibited large capacitance ranges, 100–500 pF, with relatively small changes in reverse bias: 0–5 V or 0–12 V. These newer devices allow electronically tuned AM broadcast receivers to be realized as well as a multitude of other functions requiring large capacitance changes at lower frequencies, generally below 10 MHz. Some designs of electronic security tag readers used in retail outlets require these high capacitance varicaps in their voltage-controlled oscillators.

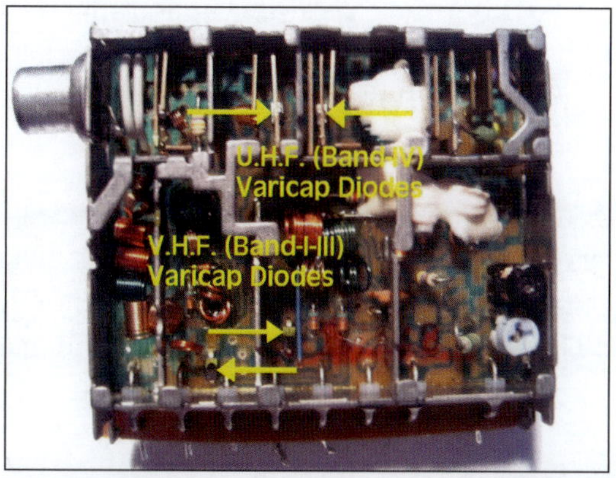

Australian market band I-III-U television tuner with varicaps highlighted.

Consumer AM-FM broadcast tuner with varicaps highlighted.

The three leaded devices are generally two common cathode connected varicaps in a single package. In the consumer AM/FM tuner depicted at the right, a single dual-package varicap diode adjusts both the passband of the tank circuit (the main station selector), and the local oscillator with a single varicap for each. This is done to keep costs down – two dual packages could have been used, one for the tank and one for the oscillator, four diodes in all, and this is what was depicted in the application data for the LA1851N AM radio chip. Two lower-capacitance dual varactors used in the FM section (which operates at a frequency about one hundred times greater) are highlighted by red arrows. In this case four diodes are used, via a dual package for the tank/bandpass filter and a dual package for the local oscillator.

Switching

Special types of varicap diode exhibiting an abrupt change in capacitance can often be found in consumer equipment such as television tuners, which are used to switch radio frequency signal paths. When in the high capacitance state, usually with low or no bias, they present a low impedance path to RF, whereas when reverse biased their capacitance abruptly decreases and their RF impedance increases. Although they are still slightly conductive to the RF path, the attenuation they introduce decreases the unwanted signal to an acceptably low level. They are often used in pairs to switch between two different RF sources such as the VHF and UHF bands in a television tuner by supplying them with complementary bias voltages.

Harmonic Multiplication

In some applications, such as harmonic multiplication, a large signal amplitude alternating voltage is applied across a varicap to deliberately vary the capacitance at signal rate to generate higher harmonics, which are extracted through filtering. If a sine wave current of sufficient amplitude is applied driven through a varicap, the resultant voltage gets “peaked” into a more triangular shape, and odd harmonics are generated.

This was one early method used to generate microwave frequencies of moderate power, 1–2 GHz at 1–5 watts, from about 20 watts at a frequency of 3–400 MHz before adequate transistors had been developed to operate at this higher frequency. This technique is still used to generate much higher frequencies, in the 100 GHz – 1 THz range, where even the fastest GaAs transistors are still inadequate.

Substitutes for Varicap Diodes

All semiconductor junction devices exhibit the effect, so they can be used as varicaps, but their characteristics will not be controlled and can vary widely between batches.

Popular makeshift varicaps include LEDs, 1N400X series rectifier diodes, Schottky rectifiers and various transistors used with their collector-base junctions reverse biased, particularly the 2N2222 and BC547. Reverse biasing the emitter-base junctions of transistors also is quite effective as long as the AC amplitude remains small. Maximum reverse bias voltage is usually between 5 and 7 Volts, before the avalanche process starts conducting. Higher-current devices with greater junction area tend to possess higher capacitance. The Philips BA 102 varicap and a common zener diode, the 1N5408, exhibit similar changes in junction capacitance, with the exception that the BA 102 possesses a *specified* set of characteristics in relation to junction capacitance (whereas the 1N5408 does not) and the “Q” of the 1N5408 is less.

Before the development of the varicap, motor driven variable capacitors or saturable-core reactors were used as electrically controllable reactances in the VCOs and filters of equipment like World War II German spectrum analyzers.

Gunn Diode

Gunn diodes have been available for many years and they form a very effective method of generating microwave signals anywhere from around 1 GHz up to frequencies of possibly 100 GHz.

Gunn diodes are also known as transferred electron devices, TED. Although is referred to as a diode, the devices does not possess a PN junction. Instead the device uses an effect known as the Gunn effect (named after the discoverer, J B Gunn).

Although the Gunn diode is normally used for generating microwave RF signals, the Gunn diode may also be used for an amplifier in what may be known as a transferred electron amplifier or TEA.

As Gunn diodes are easy to use, they form a relatively low cost method for generating microwave RF signals, often being mounted within a waveguide to form a simple resonant cavity.

Gunn Diode Symbol

There is a variety of Gunn diode symbols that may be seen used within circuit diagrams. Possibly the most widely used Gunn diode symbol uses two filled in triangles with points touching is used as shown below.

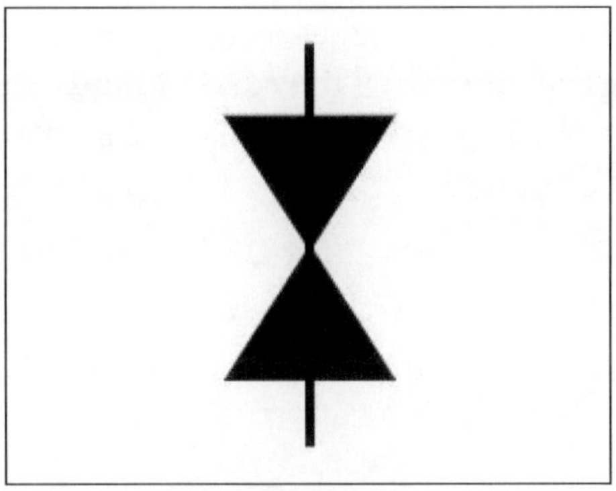

Gunn diode symbol.

Gunn Diode Basics

The Gunn diode is a unique component - even though it is called a diode, it does not contain a PN diode junction. The Gunn diode or transferred electron device can be termed a diode because it has two electrodes.

The Gunn diode operation depends on the fact that it has a voltage controlled negative resistance – this being dependent upon the fact that when a voltage is placed across the device, most of the voltage appears across the inner active region. This inner region is particularly thin and this means that the voltage gradient that exists in this region is exceedingly high.

The device exhibits a negative resistance region on its V/I curve as seen below. This negative resistance area enables the Gunn diode to amplify signals, enabling it to be used in amplifiers and oscillators. However it is the Gunn diode oscillators are the most commonly used.

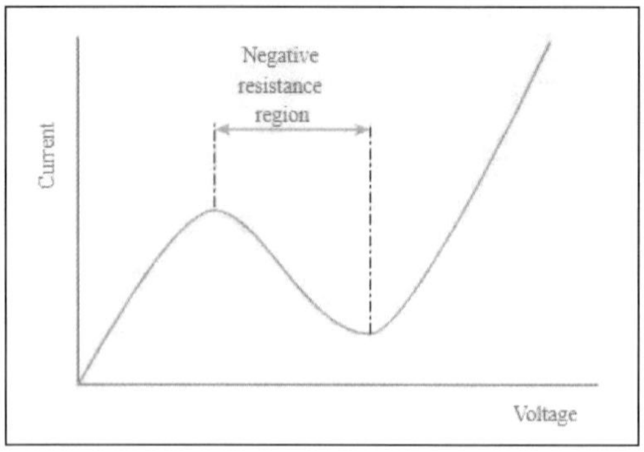

This negative resistance region means that the current flow in diode increases in the negative resistance region when the voltage falls - the inverse of the normal effect in any other positive resistance element. This phase reversal enables the Gunn diode to act as an amplifier and as an oscillator.

Gunn Diode as an Oscillator

Whilst the Gunn diode has a negative resistance region, it is interesting to see a little more about how this happens and how it acts as an oscillator.

At microwave frequencies, it is found that the dynamic action of the diode incorporates elements resulting from the thickness of the active region.

When the voltage across the active region reaches a certain point a current is initiated that travels across the active region. During the time when the current pulse is moving across the active region the potential gradient falls preventing any further pulses from forming. Only when the pulse has reached the far side of the active region will the potential gradient rise, allowing the next pulse to be created.

It can be seen that the time taken for the current pulse to traverse the active region largely determines the rate at which current pulses are generated. It is this that determines the frequency of operation.

To see how this occurs, it is necessary to look at the electron concentration across the active region. Under normal conditions the concentration of free electrons would be the same regardless of the distance across the active diode region. However a small perturbation may occur resulting from noise from the current flow, or even external noise - this form of noise will always be present and acts as the seed for the oscillation. This grows as it passes across the active region of the Gunn diode.

The increase in free electrons in one area cause the free electrons in another area to decrease forming a form of wave.

The peak will traverse across the diode under the action of the potential across the diode, and growing as it traverses the diode as a result of the negative resistance.

A clue to the reason for this unusual action can be seen if the voltage and current curves are plotted for a normal diode and a Gunn diode. For a normal diode the current increases with voltage, although the relationship is not linear. On the other hand the current for a Gunn diode starts to increase, and once a certain voltage has been reached, it starts to fall before rising again. The region where it falls is known as a negative resistance region, and this is the reason why it oscillates.

Gunn Diode Construction

Gunn diodes are fabricated from a single piece of n-type semiconductor. The most common materials are gallium Arsenide, GaAs and Indium Phosphide, InP. However other materials including Ge, CdTe, InAs, InSb, ZnSe and others have been used. The device is simply an n-type bar with n+ contacts. It is necessary to use n-type material because the transferred electron effect is only applicable to electrons and not holes found in a p-type material.

Within the device there are three main areas, which can be roughly termed the top, middle and bottom areas.

The most common method of manufacturing a Gunn diode is to grow and epitaxial layer on a degenerate n+ substrate. The active region is between a few microns and a few hundred micron thick. This active layer has a doping level between $10^{14}cm^{-3}$ and $10^{16}cm^{-3}$ - this is considerably less than that used for the top and bottom areas of the device. The thickness will vary according to the frequency required.

The top n+ layer can be deposited epitaxially or doped using ion implantation. Both top and bottom areas of the device are heavily doped to give n+ material. This provides the required high conductivity areas that are needed for the connections to the device.

Devices are normally mounted on a conducting base to which a wire connection is made. The base also acts as a heat sink which is critical for the removal of heat. The connection to the other terminal of the diode is made via a gold connection deposited onto the top surface. Gold is required because of its relative stability and high conductivity.

During manufacture there are a number of mandatory requirements for the devices to be successful - the material must be defect free and it must also have a very uniform level of doping.

Gunn Diode Advantages and Disadvantages

Like any form of component, the Gunn diode has a number of advantages and disadvantages that need to be considered when looking at suitable components for a particular circuit design.

Gunn Diode Advantages

- High bandwidth;
- High reliability;
- Low manufacturing cost;
- Fair noise performance (does not use avalanche principle);
- Relatively low operating voltage.

Gunn Diode Disadvantages

- Low efficiency below about 10 GHz;
- Poor stability – frequency varies with bias and temperature;
- FM noise high for some applications;
- Small tuning range.

References

- Diode, technology: britannica.com, Retrieved 5 March, 2019
- Hastings, Alan (2005). The Art of Analog Layout (Second ed.). Prentice Hall. ISBN 9780131464100
- Diode-working-principle-and-types-of-diode: sarutech.com, Retrieved 11 January, 2019
- "BZX79C5V6 – 5.6V, 0.5W Zener Diode - data sheet". Fairchild Semiconductor. Retrieved July 22, 2014
- "Laughton, M. A. (2003). "17. Power Semiconductor Devices". Electrical engineer's reference book. Newnes. Pp. 25–27. ISBN 978-0-7506-4637-6. Retrieved 2011-05-16
- V-i-characteristics-of-doiode: electricalstudynotes.com, Retrieved 12 July, 2019
- Sedra, Adel; Smith, Kenneth (2010). Microelectronic circuits (6th ed.). New York: Oxford University Press. P. 214. ISBN 9780195323030
- Calvert, James (15 February 2002). "Varactors". Dr Tuttle's Home Page. Retrieved 23 January 2017

Permissions

All chapters in this book are published with permission under the Creative Commons Attribution Share Alike License or equivalent. Every chapter published in this book has been scrutinized by our experts. Their significance has been extensively debated. The topics covered herein carry significant information for a comprehensive understanding. They may even be implemented as practical applications or may be referred to as a beginning point for further studies.

We would like to thank the editorial team for lending their expertise to make the book truly unique. They have played a crucial role in the development of this book. Without their invaluable contributions this book wouldn't have been possible. They have made vital efforts to compile up to date information on the varied aspects of this subject to make this book a valuable addition to the collection of many professionals and students.

This book was conceptualized with the vision of imparting up-to-date and integrated information in this field. To ensure the same, a matchless editorial board was set up. Every individual on the board went through rigorous rounds of assessment to prove their worth. After which they invested a large part of their time researching and compiling the most relevant data for our readers.

The editorial board has been involved in producing this book since its inception. They have spent rigorous hours researching and exploring the diverse topics which have resulted in the successful publishing of this book. They have passed on their knowledge of decades through this book. To expedite this challenging task, the publisher supported the team at every step. A small team of assistant editors was also appointed to further simplify the editing procedure and attain best results for the readers.

Apart from the editorial board, the designing team has also invested a significant amount of their time in understanding the subject and creating the most relevant covers. They scrutinized every image to scout for the most suitable representation of the subject and create an appropriate cover for the book.

The publishing team has been an ardent support to the editorial, designing and production team. Their endless efforts to recruit the best for this project, has resulted in the accomplishment of this book. They are a veteran in the field of academics and their pool of knowledge is as vast as their experience in printing. Their expertise and guidance has proved useful at every step. Their uncompromising quality standards have made this book an exceptional effort. Their encouragement from time to time has been an inspiration for everyone.

The publisher and the editorial board hope that this book will prove to be a valuable piece of knowledge for students, practitioners and scholars across the globe.

Index